Thinking Catherine Malabou

Thinking Catherine Malabou

Passionate Detachments

Edited by
Thomas Wormald and Isabell Dahms

ROWMAN &
LITTLEFIELD
INTERNATIONAL
London • New York

Published by Rowman & Littlefield International Ltd.
Unit A, Whitacre Mews, 26–34 Stannary Street, London SE11 4AB
www.rowmaninternational.com

Rowman & Littlefield International Ltd. is an affiliate of Rowman & Littlefield
4501 Forbes Boulevard, Suite 200, Lanham, Maryland 20706, USA
With additional offices in Boulder, New York, Toronto (Canada), and Plymouth (UK)
www.rowman.com

British Library Cataloguing in Publication Data
A catalogue record for this book is available from the British Library
ISBN: HB 978-1-78660-692-1

Library of Congress Cataloging-in-Publication Data Is Available
ISBN 978-1-78660-692-1 (cloth: alk. paper)
ISBN 978-1-78660-693-8 (electronic)

™ The paper used in this publication meets the minimum requirements of American National Standard for Information Sciences—Permanence of Paper for Printed Library Materials, ANSI/NISO Z39.48–1992.

Printed in the United States of America

Contents

Acknowledgments

The chapters and discussions collected in this volume constitute the proceedings of the first international symposium devoted to the work of Catherine Malabou.

The symposium was enabled and organized by the Centre for Advanced Research in European Philosophy (CAREP) at Kings University College in London, Ontario, Canada, as well as the Centre for the Study of Theory and Criticism at Western University in London, Ontario, Canada. Generous support for the compilation and completion of the volume was also provided by the Social Sciences and Humanities Research Council of Canada (SSHRC).

We are grateful to all the contributors for their rich work and patience in seeing this project through. The many inspiring discussions here, an ongoing dialogue with Catherine and the sponsorship and guidance of the CAREP Centre and in particular of Antonio Calcagno, made this volume possible.

Introduction

Attaching and Detaching from Philosophy with Catherine Malabou

Thomas Wormald and Isabell Dahms

Contributing to the growing body of literature that develops the work of contemporary French philosopher Catherine Malabou, this collection offers an engagement with Malabou's work through the idea of "passionate detachment." The contributors of this volume were asked to use this orienting thematic as inspiration to explore and assess Malabou's own thought and the ways her thinking uniquely contributes to the contemporary development and situation of philosophy. Each contribution represents a unique interpretation and creative exploration of what "passionate detachment" means, informed and inspired by the resources of Malabou's thought. While the approaches here are diverse in each their own respective response to thinking through the notion of "passionate detachment," they all negotiate the fact that philosophy today sees itself confronted with questions regarding its methodology, its history, and future. Philosophy cannot recuse itself in the detached purity of contemplation. It is constantly challenged by its disciplinary others—such as science, postcolonialism, feminist theory—to remember that it is firmly of and in the world. These disciplines are holding philosophy accountable to and for its past, demonstrating that philosophy is always situated, coming from a particular place, with a particular history. Philosophy is therefore asked to engage with the fact that it too, by means of its epistemological, metaphysical, and ontological accounts, institutes the economic, racial, and gendered power dynamics that furnish the world we live in.

The contributions of this volume explore in various ways how Malabou's thought both performs and creates resources for this negotiation of philosophy's attachment and detachment from other disciplines. Philosophy cannot afford to detach from either science or politics. It therefore becomes a question of how to make this attachment possible. What kind of interaction can philosophy have with either science or politics without conquering its disciplinary others? To attach and simultaneously keep a distance, but to do so passionately, as performed by Malabou, seems to be a possible way. We therefore propose to explore if and how passionate detachment can realize the stakes of politics and science. How does one carry out philosophy while subverting it, changing it, directing it on, or opening it up to different pathways? The contributions here explore the resources of Malabou's thought to perform this affective, epistemic negotiation through various acts of detachment.

The volume is then organized along the following structure. The first section explores the detachment of Malabou from her own philosophical training in deconstruction. How has Malabou transformed deconstruction and what of deconstruction remains in her thinking? The second section is on the theme of habit and asks if detachment or new attachments are possible. Habit, a central theme in Malabou's work, questions the possibility of any radical break. Does habit, the unconscious return, undercut attempts of detachment as outlined in the first chapters? The third section is entitled "Toward a Passionate Philosophy" and consists of two parts: "The Life of Science" and "The Politics of Philosophy." The first half of this section explores detachments through the relation of Malabou's thought and science. In this section, authors engage with Malabou's attempt to reconnect continental philosophy and the life sciences. The second half is devoted to thinking the passions and detachments that transpire through philosophy's confrontation with politics. Our contributors demonstrate that philosophy cannot afford to neglect the ethico-political implications of its past and future, and that philosophy must in some sense detach from its own tradition to adequately confront various pressing questions of its relation to and intrications with race, ethics, gender, colonialism, and also science.

OPENINGS

Since Malabou's work has from the start traveled in and through different languages, the volume will open with a contribution by Carolyn Shread, who responds from the position of the translator to the question of philosophy's negotiation, through attachments and detachments, from its disciplinary others. Shread does so by reflecting on the task of the translator in relation to Malabou's thought. She brings attention to the peculiarly reflexive task of translating a thinker of plasticity, metamorphosis, and transformation, drawing out some ways that Malabou's thought can help reconceptualize translation and translation studies. Shread's piece helpfully moves through various key motifs of Malabou's thought, situating and introducing Malabou to unfamiliar readers.

DETACHING FROM DERRIDA?

In her first contribution to this collection, Catherine Malabou reflects on a consonance she detects between the thought of Alain Badiou and Jacques Derrida with respect to how each thinker characterizes the relation of philosophy and their own orienting discourses of mathematics and literature. For both Badiou and Derrida, philosophy is that which responds in lieu of what Malabou describes as the silence or nonresponse of science and literature. For Badiou, science is an absolute closure that lacks nothing and thus does not respond, and, for Derrida, literature is an absolute openness, a play of the signifier, that does not need to respond, or be responsible, in its freedom. What Malabou identifies is that these two positions are speculatively identical: science as absolute closure and literature as absolute openness are two ways of expressing a nonresponsibility, a self-sufficiency that does not have to answer or respond. Malabou thus reflects on what this means for the task of philosophy, that it is charged with the responsibility to articulate the world, to force it to speak, and if we are content or want such a vision for philosophy today. This work further develops Malabou's understanding of the relationship of deconstruction, the sciences, and philosophy, and marks Malabou's first published engagement with the thought of Alain Badiou.

In the second piece to our collection, Deborah Goldgaber interrogates the extent to which Malabou has detached from deconstruction. In an important reappraisal of Jacques Derrida's *Of Grammatology*, Malabou asks, "Why, despite Derrida's claims, has a scientific 'grammatology' never seen the light of day?" On Malabou's view, the limited prospects for a "scientific" grammatology are due not to the absolute generality of writing, as Derrida claimed, but to the gramme's genetic link to schemas of inscription. In the contemporary context, she argues, the material basis of life is no longer thought in terms of code and inscription, but in terms of morphology and plasticity. Goldgaber argues, contra Malabou, that Derrida does not understand the gramme as inscribed, and this is the case because plasticity is its essential feature—in precisely the sense of "plasticity" Malabou elaborates in her work. Moreover, Goldgaber contends that plasticity itself can only be understood as an essential property of matter once we define the latter grammatologically. Challenging Malabou's claim that plasticity assumes the place of writing through a close examination of Derrida's *Of Grammatology*, Goldgaber's contribution assesses Malabou's relation to deconstruction and Derrida, while referencing the work of other thinkers indebted to Derrida, such as Claire Colebrook and Karen Barad, importantly situating Malabou within a wider strain of postdeconstructive materialisms and realisms.

In his contribution, John Nyman thoughtfully explores the detachments and transformations that occur through philosophical transmission and inheritance, assessing Malabou's relationship to Derrida by way of reexamining the way that Derrida takes up one of his formative predecessors: Edmund Husserl. Nyman reflects on the way that philosophy moves or metamorphizes through a play of manifestation and ruination, of visibility and invisibility, what Malabou characterizes as a succession of "dusks": the visibility of our predecessors' failures galvanizes the emergence of new forms of thought, simultaneously marking the decline, eclipse, or gradual invisibility of those very same predecessors. Nyman richly assesses the specificities of Malabou's own inheritance of Derrida, as well as offers striking reflections on the way in which thought itself is transmitted and transformed.

ARE NEW HABITS POSSIBLE?

Following the exploration of Malabou's detachment and attachment to aspects of deconstruction, our volume turns to the question of habits. To begin this section, Sandrine Hansen explores the fact that the phenomenon of habit has been classified as a *pharmakon*, both remedy and poison. As such, Hansen maintains, habit cannot, as some commentators have suggested, be determined merely as an ontologically adaptive phenomenon, understood univocally as an instantiation of open-ended plasticity. Whether characterized as an alienating mechanism or a compulsive force of the body, the perseverance of habit did not escape modern critics. In fact, this seemingly mechanical aspect of habit resulted in a negative evaluation of the phenomenon, downplaying, if not parenthesizing, the ontologically constitutive factor of habit in the formation of the self. But what if this "mechanism" is granted an ontologically constitutive position? What if the configuration of the self is determined or affected by habit exactly as a pharmakon, both adaptive and compulsive? To come nearer to an understanding of this elusive propensity of habit, Hansen juxtaposes Catherine Malabou's theory of compulsive repetition with Sigmund Freud's. By entering the dispute between Malabou's conceptualization of compulsive repetition as an instance of destructive plasticity and Freud's understanding of compulsive repetition as the manifestation of an inherent resistance to change characterized as organic elasticity, Hansen productively questions the uniform understanding of habit as a phenomenon of plasticity and to deepen the understanding of the ontological complexity of habit as both adaptive and obstinate. In deploying Freud against Malabou, Hansen offers a Freudian response to Malabou's incisive critique of psychoanalysis in *The New Wounded: From Neurosis to Brain Damage*.

In his piece, Cristóbal Durán explores Malabou's contribution to the development of contemporary materialism with respect to her reelaboration of the notion of form and its relationship to subjectivity, affect, and temporality. Durán first highlights the way that Malabou rescues form from its mid-century, poststructural critique by rearticulating it as possessing its own immanent capacity for transformation: whereas spirit and matter, or, in Malabou's recent language, the symbolic and the biological, are traditionally bifurcated in metaphysics, and the former term in the binary is seen as necessary for externally animating the latter, Malabou stipulates that form does not require any exterior or transcendental resources, but is capable of supporting and enacting its own self-differentiation, self-transformation, and metamorphosis. In this way, the life of form is not preformed or predetermined, but capable of a future that escapes and outstrips its own present. This possibility, however, is not strictly subject to our power—we are also subject to *it*. From this crucial insight of Malabou's philosophy, Durán examines how this capacity and vulnerability to be unexpectedly shaped or formed by a future that we did not see coming is the wellspring of a central motif in Malabou's thought, disaffection: we do not know what can or will happen and therefore we are powerless to what occurs. Yet, while this may appear to disempower us, Durán argues that disaffection can be a concept or affect through which we can conceive of alternate futures: if we are not attached to the present, and the future is not predicated on any past, who is to say our disaffection, our detachment to the past and to the present, is not the opening through which we can welcome and attach ourselves to a future that is totally unrecognizable?

In her contribution, Isabell Dahms reflects on the Hegelian concept of speculation, as it is taken up by Malabou, first in *The Future of Hegel*, where this concept opens up new possibilities of reading and writing philosophical texts, a philosophical practice that is then more specifically announced as a feminist strategy in *Changing Difference*. To explain the unfolding of speculation in Hegel's work, Dahms focuses on the *Encyclopedia of the Philosophical Sciences*, more specifically the section on habit in the "Anthropology." This dual focus on speculation and habit allows Dahms to place Hegel and Malabou in conversation. Demonstrating how Malabou's account of habit in *The Future of Hegel* both explains and problematizes Hegelian speculation, Dahms goes on to analyze how Malabou's own account of speculation transforms in *Changing Difference*. In view of this feminist turn, Dahms asks us to wonder to what extent Malabou's concept of speculation is still specifically Hegelian and could we not rather find the potential for thinking of a feminist speculation, a new habit, one that will offer alternative ways to detach from, read, navigate, and transform the philosophical canon.

In our last contribution to this section, Thomas Wormald asks if habit effectively means a being-at-home-with-oneself, if it entails a kind of having, of property or possession, a mark of the "proper," at the individual level, what are the implications of this if we transpose it to the other registers in which habit significantly functions in Hegel: at the levels of thought and world formation. No longer merely involving a coming-to-be-at-home-with-one-self, habit becomes a matter of being-at-home-with-one's-world, a demarcation of property, possession, and the "proper" with respect to the ways of being and thinking in the world. Focusing on Malabou's closing words on habit in *The Future of Hegel*, that the future depends on habitués looking at their habits, Wormald proposes an analytic of habitués and nonhabitués as a potential way to open up and think about the ethico-political implications and possibilities of Malabou's thinking.

TOWARD A PASSIONATE PHILOSOPHY: THE LIFE OF SCIENCE

Through her engagement with neuroscience and biology, Catherine Malabou has endeavored to interrupt the habits of continental philosophy. The following section sets out to analyze the stakes of this disruption, critically engaging the points of attachment between Malabou's thought and the life sciences. Joshua Schuster opens this section by asking to what extent to which Malabou's engagement with biology challenges the framework of Derrida's thought. One of Catherine Malabou's principal claims concerning Jacques Derrida and deconstruction is that Derrida did not go far enough. Specifically, Malabou refers to Derrida's concession in his later thinking of the existence of what he calls the undeconstructible, exemplified in notions such as justice and democracy. In this contribution, Joshua Schuster interrogates Malabou's critique of Derrida, concluding that we must necessarily think of some limit in order for deconstruction to not become a kind of automatic work, a calculated system guaranteed in advance. Examining the problems of extinction and biodiversity, and referencing the work of Charles Darwin, Schuster challenges Malabou's characterization of deconstruction and the adequacy of her concept of plasticity to capture and characterize biohistorical life.

In his contribution, Andrew Bevan investigates how historical attempts to explain intersubjectivity using the concepts of *Einfühlung* and empathy assume a unidirectional movement of "feeling into," from an immutable, psychical interiority to a physical "outside" that ultimately derives from an anthropocentric, dualist ontology. With the growing development of and affirmation of the phenomenon of plasticity in neurobiology, Bevan claims that these traditional accounts are no longer sustainable and that new frameworks must be thought. Adopting Malabou's formulation of brain plasticity as a simultaneous reception and donation of form, Bevan reorients the movement of empathy from its historical unilaterality in favor of a simultaneous coexistence of "from" and "to," inside and outside, in which the human simultaneously forms itself (its feelings, sensations, thinking) in its forming of this outside. Through his critique of the history of *Einfühlung*, Bevan alternately characterizes empathic encounters as affective encounters between singularities, composites, and heterogeneous entities conceived within a flat ontology of turbulent materiality.

In our next chapter, Nancy D. Nisbet draws the work of Alain Badiou and Malabou together to argue for an ontological plasticity of genetics and the possibility of evental mutation. Through her development of plasticity, Malabou opens up both the possibility of evental change in living beings and the establishment of a material form of absence through her analysis of destructive plasticity. It is through this event of the form of absence that Nisbet pursues an assessment of Malabou's thought via the work of Badiou. Beyond Malabou's insistence on the epigenetic, Nisbet mobilizes Badiou to articulate an argument for the plasticity of genetic mutation. By staging a conversation between two major French philosophical figures whose relationship, as heirs of structuralism, has not yet been fully explored or articulated, Nisbet's piece constitutes an important contribution.

THE POLITICS OF PHILOSOPHY

Our next section, "The Politics of Philosophy," articulates a number of questions often judged to be either outside of philosophy or inhabiting only its margins, for instance, questions of race, colonialism, sexuality, queer subjectivity, art, and aesthetics. Contributors in this section explore the philosophical tools Malabou provides to make philosophy "proper" respond to these challenges.

Michael Washington opens this section through a meditation on Derrida's *Glas*. Washington observes that in Derrida's reflections on Hegel and Genet in *Glas*, Derrida finds in Genet a strange force that seems to offer a way out for him of dialectical negativity in Hegel: that is to say, sexuality and more precisely, a queer sexuality, one that is enamored with annihilation. There is a unique agency at work in such a queer death drive that becomes emancipatory for Derrida, and yet, what he could not have anticipated is that this queer destructive force would also be decidedly plastic: destructive and yet paradoxically generative of something new. In this contribution, Washington pursues and attempts to elaborate this strange queer force that Derrida finds in Genet, placing him in conversation with Catherine Malabou's concept of destructive plasticity and Lee Edelman's theorization of queer negativity.

Washington fascinatingly reflects on what this means for Derrida's reading of dialectics, using Malabou and Edelman to push toward a rethinking of the possibilities of a specifically queer form of negation.

Inspired by Malabou's articulation of plasticity, in our next chapter Georgia Mouroutsou explores how the stoicism of Marcus Aurelius both resonates and enhances Malabou's thought. Mouroutsou elaborates how, if we read Malabou and Aurelius in conjunction, Malabou's vocabulary of plasticity, of giving and receiving shape or form, obtains new ethical valences and philosophical possibilities from this Stoical inflection. Aurelius urges us not to passively acknowledge unfortunate events as facts or swallow them as bitter pills, but solicits us to learn to love our fate by creatively shaping whatever comes our way, echoing Malabou's own pronouncements in *What Should We Do with Our Brain?* Mouroutsou closes her reflections with the suggestion that we start reappreciating passivity as a power. In an era that "measures" everything in accordance to what one does, makes, and produces, Mouroutsou asks whether we should not decide, instead, to appreciate the power of the mind to receive and be affected as an essential part of what makes an individual and a culture "plastic." Mouroutsou's contribution is rich, offering new inflections and interpretive directions through her establishment of Malabou's resonance with Stoic thought.

Meadhbh Mcnutt's piece takes as her starting point the question of the relationship of a self-affecting brain and art. That is, what do recent developments in neuroscience do to reshape our conceptions and affective experiences of aesthetics? Moving across the work of Malabou, Deleuze, and Spinoza, and concretizing her reflections through an examination of Lars von Trier, Mcnutt explores the interplay of neuroplasticity and affect, chaos, immanence, and composition, exploring how this particular constellation of thought offers new possibilities in the arrangement of self that arises in the convergence of plasticity, affect, and brain as a junction of planes. As a practicing artist, Mcnutt's piece is both timely and illuminating in its reflections on how our experience of art changes with new scientific understandings of the self as well as pertinent in its staging a dialogue between the work of Malabou and Deleuze.

Sujaya Dhanvantari's contribution fascinatingly explores an encounter between Frantz Fanon's mid-twentieth-century study of the colonized psyche and the current philosophical investigations by Catherine Malabou into the sociopolitical dimensions of neurological disease. Dhanvantari proposes that rereading Fanon today will entail giving his "revolutionary" psychiatric assessment of mental illness in the colonized subject another sense that blends the biological brain with its oppressive context. While Fanon will not have anticipated those concepts of cerebrality and destructive plasticity put forward by Malabou, Dhanvantari contends that we can still now find in that work an interpretation of *the neuropathological trauma of the colonized subject*. Considering some of Fanon and Malabou's substantial contributions that elaborate the link between psychic trauma and political oppression, Dhanvantari positions Fanon's critique of the negative effect of colonialism on the biological body and psyche in relation to Malabou's effacement of the border between biological and political violence in her concept of the new wounded. Tracing the unfolding of the traumatic event from the colonial to the contemporary sociopolitical state of oppression, Dhanvantari ultimately suggests that a spectral colonial trauma haunts the psychic pathologies of today's oppressed subjectivities.

CONCLUSION

Closing our collection is a second piece by Malabou, a previously untranslated essay on the notion of structure as it appears in the work of Heidegger and Claude Lévi-Strauss. Malabou stages a dialogue between the two, one that, she observes, curiously never took place. Between and through Heidegger and Lévi-Strauss, Malabou reflects on the dialectical relationship that the notion of structure entails, observing that it is at once a site of attachment and synthesis, that, nonetheless, is contingent on the play or bringing together of discontinuous particularities, detached elements. Elaborating both Heidegger and Lévi-Strauss's respective understandings of structure, Malabou herself orchestrates, between the existential analytic and structuralism, what she hears as a "strange duet" between the two. A powerful reading that both reveals and refines our understanding of Malabou's relation to both Heideggerian ontology and French structuralism, this chapter indexes Malabou's ongoing philosophical investment in transforming and passionately detaching from the traditions that have shaped her thinking. Underlying her present and future work, then, is the continuous effort to reevaluate and refine her work on the symbolic and structure, for it is this problematic that will allow her to attach and detach from philosophy's others.

It seems to us that instead of remaining attached to philosophy's self-sufficiency, its sovereignty, the work of Catherine Malabou explores possibilities of transforming, decentering and remembering, a practice or way of doing philosophy described in this volume in terms of passionate detachment. Passionate in the sense that philosophy "suffers," that it exposes itself to changes and formative influences it did not see coming and that result in dramatic alterations, as well as passionate in the sense of loving, being invested in what is to come, knowing that the adventure of thought, even and especially if it compels us to detach from our attachments, is to coparticipate in the shared sculpting of worlds of being and thought. The contributions to this volume, each in their own way, are performances of this act of passionate detachment, as they negotiate and engage a thinker to whom the contributors are indebted to and guided by, but from whom they must at the same time detach in order to obtain the critical perspective required to genuinely and reciprocally engage. For it becomes clear in Malabou's work that to remain absolutely attached, to refuse the movement and migrations of plasticity, is to remain a deluded master, one who paralyzes the movement of thought, bodies, ways of being, and knowing. Rather, Malabou's thought powerfully evinces the ethos that it is incumbent on philosophy to become passionate, to detach from itself, to "suffer" and and let itself be shaped by that which solicits or challenges it to become otherwise than it is or has been. This volume follows the ways in which Malabou demonstrates that philosophy is capable of responding, of suffering, and of undergoing transformation and change. The contributions then explore both what passionate detachment means in Malabou, offering some instances of what thinking with Malabou, and what passionately detaching can possibly open up and do.

Watching Thinking Move

Malabou in Translation

Carolyn Shread

Currently, there is considerable interest in studying translation flows in order to track the circulation of ideas globally and to better understand the shaping and impact of texts in new contexts. This chapter discusses Catherine Malabou as her work moves across multiple languages and cultures. While recognizing the importance of translation phenomena for her work, however, I seek not only to analyze the progression of her thought beyond French, but to reflect on what her philosophy says about the translation process itself. My objective, therefore, is to contribute to responses to Malabou's philosophical work by emphasizing how translation as paradigm can be read as a particularly productive instance of her articulation of plasticity.

Although Malabou herself has not written on translation as a philosophical phenomenon, in conversation she is swift to acknowledge its importance. I shall indicate some of the many ways in which these two fields of study are particularly complementary in the context of her work. Whereas for almost a decade I have sought to introduce Malabou's philosophy into the field of translation studies,[1] here my task is to draw attention to translation within the discipline of philosophy—not that philosophers are uninterested in translation. In fact, there has been an increasing convergence of the two disciplines, as exemplified, for instance, in Gayatri Chakravorty Spivak's long career as translator, translation studies scholar, and thinker. The meeting point is epitomized by Barbara Cassin's monumental *Vocabulaire européen des philosophies* (2004) and its proliferating translations, including the English *Dictionary of Untranslatables*. In Cassin's increasingly global project, philosophy is posited in the interstices between the multiple terms in different languages for any given concept. As she says, "In order to find the meaning of a word in one language, this book explores the networks to which the word belongs and seeks to understand how a network functions in one language by relating it to the networks of other languages."[2] Translated meaning here supersedes meaning as constrained by a single language, and instead of this shift representing a loss, it is construed as the wealth of philosophy.

The conference from which this chapter was initially developed was itself a translational event, bringing together an international group of scholars from Canada, the United States, the United Kingdom, Holland, Chile, and beyond to discuss Malabou's work. It is worth noting that just as the conference discussions took place largely in English, the chapters in this collection are also written in English, which is not the first language of the majority of Malabou's publications. Furthermore, it is probably fair to say that many of the conference participants, no doubt like readers of this collection, access Malabou's work primarily via translation. The fact of translation is thus central to the circulation, discussion, and development of Malabou's thought.

We should also note that Malabou is in translation in person, and we might ask at what point will she no longer be in translation, but simply become English. Although we were in bilingual Canada, she spoke in English at the conference, as she usually does in Anglophone areas of North America. This translational fact prompts me to call on her signature term by recognizing the tremendous plasticity of Catherine Malabou herself. We recall that from Université Paris VIII she became professor at Kingston University, United Kingdom, and that she frequently teaches in the United States, including a new position at University of California, Irvine. She writes increasingly in English alongside French, and her work is widely read in translation. Indeed, Malabou's plasticity is such that we might wonder if she still needs translators into English. Even so, translation is clearly integral to Malabou's overall project in philosophy and the global reception of her work. She is translated into many languages and has even more translators. By the last count, her books exist in at least nine languages in addition to French: English, Spanish, German, Italian, Dutch, Japanese, Portuguese, Dutch, and Turkish.

While Malabou may no longer need English translation, her translators, for their part, must work as a team to keep up with this prolific author. In English, I am just one of the six translators of her books, along with Lisabeth During, Steven Miller, Sebastian Rand, Peter Skafish, and David Wills. This veritable Malabou translation industry invites reflection on how her work in English participates in a well-established tradition, namely, the manufacture of French thought for and/or by Anglo-American academia. In the light of a significant French cultural policy that both promotes and funds translations, and the long-standing relationship between American scholarship and French thought, this history is one of the reasons to pause to reflect on the fact, place, and effects of translation within the context of the first international conference on Malabou's work.

My approach aligns with the work of critics studying the refiguring that is translation, thinkers who watch what happens when a text goes into translation, witnessing its transformations, alert to the ways new contexts draw out a text differently, and to the modalities of its circulation, analyzing why these movements might be necessary or desirable, and identifying those whose ends such translations serve and reflect. Aware of the many ways in which philosophers from Simone de Beauvoir to Jacques Derrida, and most recently Barbara Cassin, have been refigured in English for an American academic audience, I also seek to interrogate my own practices and purposes in the movement of Malabou's thought into English in my pen, or, rather, at my keyboard.

Let it be clear that I am not arguing that it is wrong for a text to change in translation. On the contrary, transformation is inevitable given that the nature of translation is change! Although in some instances we might wish to regulate translation through prescriptive norms or at least recognize and comment on the effects of change, it would be absurd to resist alteration *per se*. Rather than lamenting failings or oversights, lauding fidelity, or celebrating fortuitous finds, I want to be attentive to how and why texts are changed in translation.

We could start by considering the ways in which a receiving framework introduces an altered role to a text. I'll mention two brief examples. Zoology professor Howard Parshley's translation *The Second Sex* (1953) evacuated much of Simone de Beauvoir's philosophical content for reasons including, but not at all limited to, his own relative lack of familiarity with existentialism. As Anna Bogic has shown, motivated by the wish to increase sales, Knopf publishers was equally involved, if not largely responsible, for this reframing of de Beauvoir as the writer of a female sex manual parallel to the Kinsey Report, rather than as a philosopher.[3] Clearly, *Le Deuxième Sexe* (1949) was strongly affected by its translation into English, for instance, in the subsequent translations that were made from the English translation, rather than the French text. Another example of transformed philosophical frameworks is Claire Goldberg Moses's presentation of the argument that "French Feminism" was in fact the creation of American academia and that the high theory trinity of Cixous, Kristeva, and Irigaray was not an accurate reflection of the variety of feminisms in France at that period, notably the activist commitments and activities of the *Mouvement de libération des femmes*.[4] Again, translation is selective and interested, repositioning texts in new contexts for new purposes. In the case of Malabou, it is critical to acknowledge this heritage and to consider how her texts in English participate or deviate from the effects of the preexisting reputation of French philosophy—and specifically women's writing—in the United States.[5] Moreover, there is work to be done in identifying the agenda Malabou's work supports, knowingly or unknowingly, in English language contexts. How is it read differently, for instance, in Australia, Great Britain, or the United States?

We might also consider the ways that an author can use translation into multiple languages, and particularly powerful ones, notably the currently hegemonic English, in order to heighten their reception not only in new spaces, but also in their original context through what we might call a boomerang translation effect. This argument could be made for the reception of Malabou in France following her global translations, although even as I make this claim I am alert to thinking about how Malabou's instinct for, and attraction to, resistance—one of the key features inherent to plasticity—might play out in this seeming acknowledgment and use of the power and influence of English. And then too I might wonder: has either a resistance to, or collaboration with, power manifested in my own, or in other, translations of her work into English? What do we do not only with Malabou's texts, but also with the language we are translating into, namely, English?

I am sensitive to participating in a tradition that has imported French philosophers into a US academic context, not because such importation and the associated status is wrong, but because it is important to recognize and study this process as a translation effect, rather than treating it as transparent communication or an unproblematic or invisible aspect of globalization. I particularly appreciate the attention to such effects shown in the commentary of Peter Skafish, translator of Malabou's *The Heidegger Change: On the Fantastic in Philosophy* (2011).[6] In his preface, Skafish reflects on the style and language he used in the translation within the specific history of translations of twentieth-century French theory. Skafish comments that by contrast to an earlier tradition, he chose to use "an English disencumbered of heavy Gallicisms" and "a less estranged English" (xiii). Beyond the clarity that I so appreciate as one of the hallmarks of Malabou's thought and writing, Skafish explains that this language—so markedly different from a tradition associated with difficult French theory—is prompted by a wider cultural framework that reflects new political and socioeconomic realities, notably "the recently '(ex)changed' statuses of French and American thought" (xvii). He goes on to argue that "a phase of obeisance to French literature, human science, and philosophy is no longer the almost obligatory aspect of an intellectual formation in the United States that it was even ten or fifteen years ago . . . the situation is thus reversed, what is left of French thought needs to be rearticulated in an idiom that . . . is far closer to . . . a generally American English" (xvii). Skafish's stylistic choices in translating Malabou, reflecting and responding to the norms of this particular moment in time, are but one instance of the way we can watch thinking move, slip into new forms, new idioms, exercising the plasticity of text in translation.

In my own experience, to translate Malabou is to engage with a striking and peculiar reflexivity since her philosophy of plasticity is so entirely apt for talking about translation and has proven to be highly productive in allowing me to dislodge certain stubborn cultural and patriarchal prejudices that keep translation fettered to the outdated paradigms underlying conventional communication and theological models of translation. In a word, the generalized impulse to regulate and restrict translation by subjugating it to an original text derives from predominant masculine models of authorship, authority, and intersubjectivity. Plasticity is a philosophical tool that helps unseat such prejudicial norms, and translation a critical and literary practice that, despite everything, never ceases to manifest the plasticity that is typically constrained or discredited.

My encounter with Malabou began with a short article that inspired the title for this chapter. For the February 2007 issue of the journal *Communication Theory*, I translated *Un œil au bord du discours* as "An Eye at the Edge of Discourse."[7] In this beautiful text, Malabou asks, "What is it to *see* a thought? To see a thought coming?" As her translator, my question, in parallel with this investigation, is to ask, as I translate, what is it to watch thought move through language into new cultural and intellectual landscapes? What is it to watch, watch over, guide, and—one of my special joys—work with the author as her thinking moves from French to English? Unlike the arrival of a new thought, there is already something discernable in the source text, but when does it take shape in translation? And how does that figure reappear, differently? Furthermore, as I highlighted above, due to the specific nature of her philosophic intervention, to translate Malabou is to simultaneously engage textually in a practice of her philosophy, to put to the test, embody, and give form to the very plasticity she presents.

In seeking to bring together translation studies and philosophy, I shall summarize the ways in which, in addition to transforming in the new languages of her work, Malabou's thought has contributed to my own articulations of translation theory. My brief translators' forewords to four of her books offer a series of statements rethinking the nature of translation, focusing above all on formulating the relationship that exists between a source text and its translations, their mutual plasticity, and this practice as the expression of hospitality.

It was first while translating *Plasticity at the Dusk of Writing: Dialectic, Destruction, Deconstruction*[8] that I was struck by the self-reflexivity of the translating process with regard to Malabou's philosophical interventions. In this instance, I took the contrast she establishes between elasticity and plasticity as a tool to define two paradigms of translation. First, elastic translation—the conventional, still dominant view, mired in a debate that asks endlessly, "is this a literal, word-for-word translation, is it an adaptation, a free translation, or an abusive translation?" It is mired here because this paradigm does not, in fact, allow translation to take place or exist given that in an elastic model premised on malleability and flexibility, rather than plasticity, translation must always revert back to, measure itself against, an original form. Translation is then forever tied up in the impossible exchange of equivalents, substitutions, conversions, incapable of recognizing different currencies, or a being that moves, a transformational ontology. By contrast, the alternative plastic translation paradigm accounts for both the giving and resistance of the source text and translation, and their mutual negotiation of hospitality. After translating this first book, I adopted the signature, *traductrice, plasticienne textuelle*, as a mantle to assert the plastic rights of translation. I then went on to muse further on these notions, developing them over the course of the following three translations.

In *Changing Difference: The Feminine and the Question of Philosophy*,[9] as I worked, the idle part of my mind explored what Malabou had to say about changing change, and thereby, changing the change that is translation. I found here again my instinctive understanding of translation as a generative activity, rather than one that reduces the text, or even resuscitates it or offers it a means to survival, as Derrida would have it ("the task of the translator is precisely to respond to this demand for survival which is the very structure of the original text").[10] This alliance with a feminine relation to generative change echoed earlier arguments I had made for metramorphic translation, based on psychoanalyst Bracha Lichtenberg Ettinger's writings.[11] A metramorphic, rather than metaphoric, model would be premised on a severality that avoids the dichotomy of text/translation construed as separate and discrete entities. Instead, the translation grows with the text, not depleting it but reusing and recycling it.

Malabou's *Ontology of the Accident: An Essay on Destructive Plasticity*[12] led me to consider accident as the condition of possibility of translation. Viewing destruction as the obverse of plasticity's power to yield to and take form is a powerful tool of analysis for translation, so long associated with deteriorating and deviating activities. To see the very ontology of translation as accident is ironically liberating, given a history of recrimination precisely for the accidents that are translation. As I explained earlier, a plastic translation paradigm, that is, one that assumes its destruction as much as its formative abilities, moves beyond the difference that is but an extension of identity to meet the possibility of its accident. This new paradigm is premised not on being, but on becoming. Indeed, how could translation be anything but becoming? And yet for so long, translation has been viewed as secondary precisely for not just being.

What type of becoming is presented then in this new translational paradigm? My fourth translation, Malabou's *Before Tomorrow: Epigenesis and Rationality*, helped me spell out the becoming of translation as an epigenesis of the text. This conception combines generative forces with the necessary destruction that is also inherent to plasticity. My conclusion, while translating Malabou's latest, is that we have been thinking translation solely in genetic terms and therefore failing to account for, or recognize, epigenetic processes. In her bold and innovative arguments in *Before Tomorrow*, Malabou claims that in Kant rationality—the transcendental itself—is subject to epigenesis. In the phrase "as it were a system of epigenesis of pure reason" in §27 of the *Critique of Pure Reason*, she finds a new key to the development of critique itself: epigenesis. For the sake of clarity, we might ask, what is epigenesis? In biology, epigenesis is the process of cellular differentiation. For a short-hand description, we could say that while genetics is concerned with the DNA sequence, epigenetics refers to gene expression. Or, to cite Malabou's explanation: "Epigenetics studies the mechanisms that modify the function of genes by activating or deactivating them. *In so far as these modifications never alter the DNA sequence itself, epigenetics is said to work on the 'surface' (epi—) of the molecule*."[13] My suggestion is that we posit translation as an epigenesis of the text, one that works precisely at the surface of the text until, as Malabou writes, "their difference disappears right into their contact."[14] Epigenesis helps us think through difference in translation outside of a binary relation. Instead, we take account of the tectonic shifts produced by translational encounters.

In asserting the productive possibilities of this new insight, I should address an issue that might be offered as a criticism of my proposals: aren't you simply using Malabou's philosophical terms as analogies for translation? Aren't plasticity and epigenesis effectively just new metaphors, adding to the many that exist? Certainly translation, perhaps because it is a relational art, has generated no end of metaphors. But again Malabou is of assistance in that she too addresses this critique. At a key turning point in *Before Tomorrow*, she writes, "If it turns out that epigenesis is only an image with nothing other than an exoteric, pedagogic or illustrative role, then my entire elaboration is meaningless."[15] Malabou acknowledges that the "as it were" in §27 would seem to point to comparative thought. But her argument is, in fact, that epigenesis is constitutive: a scientific claim. So while the field of translation is rife with metaphors, epigenesis of the text is not an analogy but rather a descriptor, beyond illustration: not as it is, but what is. Translation then, as the manifestation of textual epigenesis, instantiates the plastic life of the text.

There is much more to be said about translation plasticity, especially mining the ideas presented in *Before Tomorrow* in regard to the epigenetic process of translation. However, instead, I'll close this chapter, this stage in my parsing of the translational ramifications of Malabou's philosophy of plasticity, by emphasizing the tremendous energy produced by translating this body of work. As this brief overview indicates, every translation has brought with it new understandings of that very material engagement with text that is translation, replacing outdated conceptions of a practice that has so often been derided for precisely that engagement with materiality. Every instance where I watched, and watched over, as Malabou's thought moved into English, generated a be-coming in translation. Not every translator is as lucky to learn about what they are doing as they do it. To be given new tools, every time they come to work. But Malabou's philosophy brings to translation studies—as to so many fields—the new motor scheme of plasticity that speaks to our time and helps interpret our practices.

NOTES

1. See Carolyn Shread, "The Horror of Translation," Special Issue: "Plastique: Dynamics of Catherine Malabou," *Theory @ Buffalo* 16 (2012): 77–95 and "Catherine Malabou's Plasticity in Translation," *Traduction, terminologie, rédaction* 24, no. 1 (2012): 125–48.

2. Barbara Cassin, *Vocabulaire européen des philosophies: Le dictionnaire des intraduisibles* (Paris: Le Seuil/Le Robert, 2004). *Dictionary of Untranslatables: A Philosophical Lexicon*, trans. Emily Apter, Jacques Lezra, and Michael Wood (Princeton, NJ: Princeton University Press, 2014), xvii.

3. Anna Bogic, "Uncovering the Hidden Actors with the Help of Latour: The 'Making' of *The Second Sex*," *MonTI* 2 (2010): 173–92.

4. Claire Goldberg Moses, "Made in America: 'French Feminism' in Academia," in *Beyond French Feminisms: Debates on Women, Politics, and Culture in France, 1981–2001*, ed. Roger Célestin et al. (New York: Palgrave Macmillan, 2003), 261–84.

5. It should be noted that many of Malabou's books in English—certainly most of those that I have translated—have been published by Polity Press, which was founded in the United Kingdom. This fact complicates the narrative about the traveling of her thought into English to the United States by introducing the United Kingdom as a mediating space. Moreover, her teaching position at Kingston University in London highlights the role of the United Kingdom as an initial context for bringing her work into an Anglophone sphere.

6. Catherine Malabou, *Le Change Heidegger. Du fantastique en philosophie* (Paris: Éditions Léo Scheer, 2004). *The Heidegger Change: On the Fantastic in Philosophy*, trans. Peter Skafish (Albany: SUNY Press, 2011).

7. Catherine Malabou, "Un Œil au bord du discours," *Études phénoménologiques* XVI, no. 21–32 (2002): 209–22. "An Eye at the Edge of Discourse," trans. Carolyn Shread, *Communication Theory* 17, no. 1 (2007): 16–25.

8. Catherine Malabou, *La plasticité au soir de l'écriture: Dialectique, destruction, déconstruction* (Paris: Léo Scheer, 2005). *Plasticity at the Dusk of Writing: Dialectic, Destruction, Deconstruction*, trans. Carolyn Shread (New York: Columbia University Press, 2009).

9. Catherine Malabou, *Changer de différence: Le féminin et la question philosophique* (Paris: Galilée, 2009). *Changing Difference: The Feminine and the Question of Philosophy*, trans. Carolyn Shread (Cambridge: Polity Press, 2011).

10. Jacques Derrida, *The Ear of the Other: Otobiography, Transference, Translation*, ed. Christie McDonald, trans. Peggy Kamuf and Avita Ronell (Lincoln: University of Nebraska Press, 1985), 122.

11. See Carolyn Shread, "Metamorphosis or Metramorphosis? Towards a Feminist Ethics of Difference in Translation," *TTR* 20, no. 2 (2008): 213–42, and Luise von Flotow and Carolyn Shread, "Metramorphosis in Translation: Refiguring the Intimacy of Translation beyond the Metaphysics of Loss," *Signs: Symposium on Translation, Gender and the Hegemony of English* 39, no. 3 (2014): 592–96.

12. Catherine Malabou, *Ontologie de l'accident: Essai sur la plasticité destructrice* (Paris: Léo Scheer, 2009). *Ontology of the Accident: An Essay on Destructive Plasticity*, trans. Carolyn Shread (Cambridge: Polity Press, 2012).

13. Catherine Malabou, *Avant Demain: Épigenèse et rationalité* (Paris: Presses Universitaires de France, 2014). *Before Tomorrow: Epigenesis and Rationality*, trans. Carolyn Shread (Cambridge: Polity Press, 2016), 79.

14. Malabou, *Before Tomorrow*, 157.

15. Malabou, *Before Tomorrow*, 181.

BIBLIOGRAPHY

Bogic, Anna. "Uncovering the Hidden Actors with the Help of Latour: The 'Making' of *The Second Sex*." *MonTI* 2 (2010): 173–92.

Cassin, Barbara. *Vocabulaire européen des philosophies: Le dictionnaire des intraduisibles*. Paris: Le Seuil, Le Robert, 2004. *Dictionary of Untranslatables: A Philosophical Lexicon*. Translated by Emily Apter, Jacques Lezra, and Michael Wood. Princeton, NJ: Princeton University Press, 2014.

Derrida, Jacques. *The Ear of the Other: Otobiography, Transference, Translation*. Edited by Christie McDonald. Translated by Peggy Kamuf and Avita Ronell. Lincoln: University of Nebraska Press, 1985.

Goldberg Moses, Claire. "Made in America: 'French Feminism' in Academia." *Beyond French Feminisms: Debates on Women, Politics, and Culture in France, 1981–2001*. Edited by Roger Célestin et al., 261–84. New York: Palgrave Macmillan, 2003.

Malabou, Catherine. *Avant Demain: Épigenèse et rationalité*. Paris: Presses Universitaires de France, 2014. *Before Tomorrow: Epigenesis and Rationality*. Translated by Carolyn Shread. Cambridge: Polity Press, 2016.

———. *Changer de différence: Le féminin et la question philosophique*. Paris: Galilée, 2009. *Changing Difference: The Feminine and the Question of Philosophy*. Translated by Carolyn Shread. Cambridge: Polity Press, 2011.

———. *Ontologie de l'accident: Essai sur la plasticité destructrice*. Paris: Léo Scheer, 2009. *Ontology of the Accident: An Essay on Destructive Plasticity*. Translated by Carolyn Shread. Cambridge: Polity Press, 2012.

———. *La plasticité au soir de l'écriture: Dialectique, destruction, déconstruction*. Paris: Léo Scheer, 2005. *Plasticity at the Dusk of Writing: Dialectic, Destruction, Deconstruction*. Translated by Carolyn Shread. New York: Columbia University Press, 2009.

———. *Le Change Heidegger: Du fantastique en philosophie*. Paris: Éditions Léo Scheer, 2004. *The Heidegger Change: On the Fantastic in Philosophy*. Translated by Peter Skafish. Albany, NY: SUNY Press, 2011.

———. "Un Œil au bord du discours." *Études phénoménologiques* XVI (2000): 209–22. "An Eye at the Edge of Discourse." Translated by Carolyn Shread. *Communication Theory* 17, no. 1 (2007): 16–25.

Shread, Carolyn. "The Horror of Translation." Special Issue: "Plastique: Dynamics of Catherine Malabou." *Theory @ Buffalo* 16 (2012): 77–95.

———. "Catherine Malabou's Plasticity in Translation." *Traduction, terminologie, redaction* 24, no. 1 (2012): 125–48.

———. "Metamorphosis or Metramorphosis? Towards a Feminist Ethics of Difference in Translation." *TTR* 20, no. 2 (2008): 213–42.

von Flotow, Luise, and Carolyn Shread. "Metramorphosis in Translation: Refiguring the Intimacy of Translation beyond the Metaphysics of Loss." *Signs: Symposium on Translation, Gender and the Hegemony of English* 39, no. 3 (2014): 592–96.

Section 1

DETACHING FROM DERRIDA?

The Future of Deconstruction after Malabou

Chapter 1

Is Science the Subject of Philosophy? Miller, Badiou, and Derrida Respond[1]

Catherine Malabou

Translated by William Samson

I wish to make you aware of two reading experiences I have had. The first is that of reading Alain Badiou's 1969 text, published in *Cahiers pour l'analyse*, titled "Mark and Lack," where we can find the affirmation that gave rise to my title: "Science is the subject of philosophy."[2] This chapter is a response to Jacques Alain Miller's "Suture: Elements of the Logic of the Signifier,"[3] published a few years prior to Badiou's essay in the same journal. Despite their specific and very dated context—their questions about science had to confront, in that time, both psychoanalysis and Marxism—they contain very important elements for thinking the relations between science and philosophy today, and they have import that, in my opinion, goes well beyond their time.

The second discovery is altogether different. While, for other reasons, I was recently rereading Derrida's "Passions: An Oblique Offering,"[4] I was struck by the coincidence—if not the similarity—of the analyses developed in that book with those developed by Badiou about the *situation* of philosophy. The word "situation" is being taken here in its proper sense, as venue, place, and orientation, all at the same time. This coincidence is more surprising given that the two thinkers have little in common with each other, as is attested to by the extreme difference in the points of departure of their respective discourses. Badiou undertakes the analysis of the relations between philosophy and mathematical logic, while Derrida concerns himself with the relations between philosophy and literature. Despite it all, their conclusions strangely and mutually echo each other.

In both cases, what I will call the law of the fault of philosophy (*la loi du défaut de philosophie*) emerges. For both thinkers, philosophy is in need of an answer. It wants to answer for everything, even, which is the problem, for that which does not respond. What does not respond, for Badiou, is science. For Derrida, literature. Examining these two types of nonresponses more closely will constitute my subject, which is precisely the question of the subject. There is no subject of science, says Badiou; there is no subject of literature, says Derrida. This double absence of subject constitutes precisely, for them, the subject of philosophy, which is supposed to respond in their place by wrongly characterizing this absence as a lack. A lack which it is possible not to fill, but at least to make speak.

* * *

Let us start with Miller. He undertakes to show that science, and in particular mathematical logic, proceeds from a denial of lack, and sutures closed all that could leave the place of the other of science (which is to say, the subject-function) empty. By definition, science is "objective" and admits no lacuna or void, no desire nor gap. This constitutes the "suture" or foreclosure that Miller, with the help of Lacan, intends to deconstruct. To this end, Miller begins with a reading of Frege's *The Foundations of Arithmetic*, where he defends the idea that zero, in mathematics, is precisely not a void or a lack.[5] Miller will then demonstrate that zero in reality marks the place of the subject's foreclosure.

In *The Foundations of Arithmetic*, Frege defines number. A number refers to no particular thing. It does, however, have an object. What is the difference between a thing—that empirical X—and an object? The difference between four balls as things and what the number 4 measures? What do things become once they are numbered and numerable, in other words? They become *units*. Because of this, they obey particular syntactic determinations, which are ordered by a fundamental rule or structure, which Frege outlines in the following way: Numbers are extensions of concepts. The number 4, for example, is the extension of the concept "four." The two are equinumerical. "The number which belongs to the concept F," says Frege, "is the extension of the concept 'equinumerical to the concept F.' "[6]

What does "concept" mean here? "Concept" signifies self-identity. Saying that a number is an extension of a concept signifies that a number is identical to its concept, that it is identical to itself. In other words, as Miller rightly says, all numbers presuppose "the concept of identity to a concept."[7] This rule of identity to a concept is valid for all numbers. "The concept of identity to a concept" works for every number. This is the rule of unity: being self-identical means to "be one." There is thus some "one" in all numbers. "This one [that of the singular unit] . . . is common to all numbers in so far as they are first constituted as units."[8]

The object is the thing become one (self-identical), and thus numberable. "That definition, pivotal to his concept," continues Miller, "is one that Frege borrows from Leibniz. It is contained in this statement: *eadem sunt quorum unum potest substitui alteri salva veritate*. Those things are identical of which one can be substituted for the other *salva veritate* without loss of truth."[9] Numbers are substitutable for one another insofar as in them, at any time, the self-identical repeats itself, that is, the "one" of the unit. We can thus "pass" from one number to another without losing truth, since identity is preserved.

But the problem then arises of knowing how one "progresses" from one number to another. How the number can "pass from the repetition of the 1 of the identical to its ordered succession: 1, 2, 3, 4, 5?" The answer is for that to occur, "the zero has to appear."[10] And Miller here makes reference to the very influential Fregian analysis of zero.

Why is the "engendering of the zero"[11] necessary for Frege? At this point, we've posited the self-identity of the concept of number, and thus of the number itself. That's to say that "non-identity with itself is contradictory to the very dimension of truth." How then to designate nonidentity? "To its concept, we assign the zero," says Miller, reading Frege. Nothing falls under the concept "nonidentical with itself" if not, Miller continues, a void or a gap. Zero is the name of that nothing. Yet Frege defines zero as a number, to which he assigns the cardinal 1. This is where the "suture" comes from, from that eclipse of the zero by the one.

The non–self-identical answers fully, for Frege, to the principle by which, for any object, it must be possible to say under what concept it must be subsumed. Yet zero can be subsumed under the concept of non-self-identity as "identical to zero." Thus, zero is identical to its concept, the concept of non-self-identity. Let there be a concept "identical to zero." One and only one object, zero, is subsumed under that concept. "One" is thus by definition the cardinal number that belongs to that concept. Zero is "one" in the sense of self-identity. Now, in trying to make the number 1 appear, it must be shown that something can immediately follow zero in the series of natural numbers. "Zero," Miller says, following Frege, "is the number assigned to the non-self-identical. This number is 1 [marked zero]. It follows that 1 follows immediately from 0 in the series of natural numbers."[12] And from there we can deduce the set of numbers, following the structure of "following from," with the restriction that no number can follow itself in the natural series of numbers beginning with zero. Inscribing zero as a point of departure for the series of numbers makes it possible that the rule n + 1: 0 (self-identity of the concept "not identical to itself") is followed by 1, which itself is followed by 2, then by 3, and so on.

Examining the Fregean argument, Miller concludes that the zero is at the same time summoned and dismissed. "That which in the real is pure and simple absence finds itself through the fact of number (through the instance of truth) noted 0 and counted for 1."[13] The word "suture" signifies (1) the uniting, by use of thread, of divided parts after an accident or surgical intervention and (2) the immobile articulation characterized by two jointed surfaces united by fibrous tissue, like those which from the cranium, the apparent line constituting the conjunction of two parts. The suture is thus always at the same time a division and conjunction. In every case, the seam, the conjunction, remains visible.

What is really sutured in Frege's discourse in particular and in formal scientific language in general? Miller responds: the subject-function. "To designate it I choose the name of suture. Suture names the relation of the subject to the chain of its discourse."[14] The appearance-disappearance of the zero in the series of numbers figures the appearance-disappearance of the subject in the chain of its discourse.

Frege affirms, however, multiple times that logic does not follow from a subjective act, which is always psychological. Yet it is precisely this which appears to Miller as a denial. According to Frege, the subject counts for nothing. From this, it follows that suturing the zero can be read as suturing the subject. Indeed, for Miller, only the subject-function can subtend the operations of abstraction, of unification, and of progression. There is, therefore, identity only for a subject. It is the subject that produces the primary unity; it is impossible to think self-identity outside of the subject, since the subject-function *is* the identity-function. We've known since Descartes that a subject is, by definition, a power identical to itself. The subject is the form of identity. At the same time, like the zero, the subject is never self-identical. Like the zero, it receives its identity from a lack, it misses itself. The relationship of the zero to the chain of numbers is the same as the relationship of the subject to the chain of discourse. Like the zero, the subject is both present and absent at once. "It figures [in the chain] as the element which is lacking, in the form of a stand-in. For, while there lacking, it is not purely and simply absent."[15] Lacan shows that, in the same way that the zero is excluded from the beginning from the chain of numbers, the subject is excluded from the field of the Other, which is what comes to bar the subject. The subject displaces this bar onto the A, "a displacement whose effect is the emergence of signification signified to the subject."[16] Yet "untouched by the exchange of the bar, this exteriority of the subject to the Other is maintained, which institutes the unconscious."[17] The summation of the subject in the field of the other calls for its annulment.[18] The subject, in this way, is always alienated from and by the very process of its signification.

Regarding Frege, we have spoken of (1) unity and (2) the role of 0 in succession, its status as number, and finally of its eclipse by the 1. This structure of appearance-disappearance, of suture, would be the point, emergent or derived, of a more originary logic which Miller, with Lacan's help, proposes to name the "logic of the signifier"—a logic that proposes to "make itself known as a logic at the origin of logic,"[19] which retraces the steps of logic, a "retroaction,"[20] or a repression. "What is it that functions in the series of whole natural numbers to which we can assign their progression? And the answer, which I shall give at once before establishing it: in the process of the constitution of the series, in the genesis of progression, the function of the subject, miscognized, is operative."[21] Zero is the placeholder of the subject, which is itself, insofar as it is sutured, a placeholder.

Analyzing more closely the function and erasure of the zero, Miller distinguishes two axes. A vertical axis: the zero marks the bar on truth, separating the non–self-identical from the self-identical; insofar as it is a unit, it delimits a field. At the same time, he finds a horizontal axis: it erases that bar, since it represents itself in this field as "subsequently cancelling out as meaning in each of the . . . numbers which are caught up in the . . . chain of successional progression."[22] In some sense, the zero becomes something like a silent letter present in every number, insofar as they are self-identical but "nonidentical" to other numbers. Miller concludes, "The impossible object, which the discourse of logic summons as the not-identical with itself and then rejects as the pure negative, which it summons and rejects in order to constitute itself as that which it is, which it summons and rejects wanting to know nothing of it, we name this object, in so far as it functions as the excess which operates in the series of numbers, the subject."[23]

* * *

In a gesture the appropriateness of which I will not here question, Badiou assimilates psychoanalysis and philosophy, both being effects, according to him, of "ideology," or, rather, constituting it. Criticizing Miller's argument, he lays into the logic of the signifier, which is to say he lays into the psychoanalytic viewpoint, and slides into philosophy, exaggerating the conclusions of his criticism of the latter with regard to science. This exaggeration makes up the object of section 4 of his article, titled "Torment of Philosophy."[24]

To synopsize the salient points of Badiou's argument, he attempts to show that the concept of lack is profoundly alien to science, that science lacks nothing. It is in this sense that science does not respond; has nothing to answer for itself; and does not need to justify any denial, any foreclosure, or any suture. The torment of philosophy, which echoes that of psychoanalysis, comes precisely from the fact that it thinks of itself as the answer to science, as that which gives science its subject—a subject which science has little use for. Here, Badiou is essentially rejecting the concept of alienation. If we identify the zero-function with the subject-function, as Miller does, and if we say that the structure of the subject comes from its alienation, which is to say its subjection to the Other, then we make the discourse of logic itself an alienated discourse, thereby giving psychoanalysis priority over logic. Only psychoanalysis would allow one to pass from alienated truth to the truth of alienation, which would account for its superiority over logic.

In "Mark and Lack," the stakes of which we can now begin to situate, Badiou lashes out at Frege and Miller at once. "In our view, *both* Frege's ideological representation of his own enterprise *and* the recapture of this representation in the lexicon of signifier, lack and the place of lack, mask the pure productive essence, the positional process through which logic, as machine, lacks nothing it does not produce elsewhere."[25] It is thus necessary to construct something to oppose the discourse of lack.

And here it is: "To this twofold process (preservation of the true; convocation and marking of lack), we will oppose the stratification of the scientific signifier." Starting from logic, Badiou will later extend his concept of stratification to mathematics and physics. What *is* stratification, here? Badiou elaborates, "The theory of logic pertains to the modes of production of a division in linear writing." Logic thus gains its authority from a cut in language between signifying signs—the signs of language—and pure signifiers without signifieds, which are its own marks and conventions.

Logic begins with a stock of graphic marks, the alphabet, *a*, *b*, *c*, . . . which it cuts from their signifying milieu. It thus constitutes a collection of traces. It is this cut that Badiou opposes to the suture, a suture which the zero (or non–self-identity) is meant to introduce in the chain of numbers.[26] As opposed to the suture, which, as we have said, always shows the mark or the scar of what it cuts across, the cut in language effectuated by logic is irrevocable and unambiguous. If logic is inderivable and undecidable, Gödel's theorem shows, it is not because there is a vague zone between normal language and logic, the place of a suture. The cut is perfectly clean, leaving out nothing and no one, which "presupposes the existence of a dichotomic mechanism that leaves no remainder," which is to say "an autonomous order that is indeed closed, which is to say, entirely decidable."

Logic is a system of traces, closed in upon itself. Consequently, any concept of a tear or of a suture can be nothing more than a "psychical" effect of this closure, *it takes shape only in ideology* (which is to say in psychoanalysis and philosophy). The suture is an *a posteriori* effect, "the inevitable price of that closure" for the disciplines or discourses that suffer it, whereas such a suffering is alien to science. "What must be said, instead, is that the existence of an infallible closed mechanism conditions the existence of a mechanism that can be said to be unclosable, and therefore internally limited. The exhibition of a suture presupposes the existence of a foreclosure." The relations are inverted now. By revealing lack as the original truth, psychoanalysis and philosophy deduce or derive it, aiming to trace their own lack of closure and independence onto science.

Badiou is playing with the word "foreclosure"—the foreclosure of science is not a psychotic foreclosure, only ideology would propose such an interpretation, by struggling to characterize the scientific discourse as a discourse alienated from the fact of its presupposed lack of subject. Badiou continues, "Gödel's theorem is not the site of separation's failure, but of its greatest efficacy."[27] In the same way, "The zero is not the mark of lack in a system, but the sign that abbreviates the lack of a mark." If there is lack, it does not even appear, it puts an end to itself, immediately compensated for by a substitution. The signs used in logic are mutually substitutable; they can replace each other according to the rules of their closed system. And there we find the "identity of the signifying order with itself."[28] "The logico-mathematical signifier is sutured only to itself—it is indefinitely stratified."[29] There is no substantial self-identity. Here, identity is that of marks, traces, by definition without external model or referent, which can replace each other, which are interchangeable, all without creating holes in presence.

At this point in the analysis, we see the expansion of this stratification to sciences other than mathematical logic: "Moreover, it is science as a whole that takes self-identity to be a predicate of marks rather than of the object. The rule certainly holds for the facts of writing proper to mathematics. But it also holds for the inscriptions of energy proper to physics."[30]

If we define self-identity by substitutability—if, in other words, we move from essence to sign—then, because nothing is not substitutable-for-itself, there is no longer any "non–self-identical." And, in any case, the nonidentical is nonetheless substitutable, which is proven by the substitution of the 0 by the 1. What's not thinkable for mathematics is that which cannot be substituted for itself. But this "unthinkable" does not constitute an "unthought (*un impensé*)." Science does not have an unthought, and this is what makes philosophy and psychoanalysis suffer. We could just as well say that this is a radical unthought, if we understand by this that it is not part of the order of scientific thought: "What is not substitutable-for-itself is something radically unthought, of which the logical mechanism bears no trace. It is impossible to turn it into an evanescence. . . . What is not substitutable-for-itself is foreclosed without appeal or mark."[31]

Subsequently, scientific writing is a "play of appearances and disappearances between successive signifying orders; never exposed to the convocation of a lack, whether in the object or the thing."[32] Later, he continues, "On this side of the signifying chain, if the latter is scientific, there are nothing but other chains. If the signifier is sutured, it is only to itself. It is only itself that it lacks at each of its levels: it regulates its lacks without taking leave of itself. The scientific signifier is neither sutured nor split, but stratified."[33] We must remember that for Miller, the suture is a "placeholder" for the subject. Now, according to Badiou, because there is no suture in the signifying chains of logic, *there is no subject of science*.

"Must we therefore renounce [*annuler*] the concept of suture?" asks Badiou—must we therefore renounce the subject-function? No. It is rather a matter of "assigning to it its proper domain." Its domain is ideology, and Badiou now slides from psychoanalysis into philosophy, which makes this state of affairs perfectly obvious: "We will call 'philosophy,'" he says, "the ideological region specializing in science."[34] Philosophy is "charged with effacing the break by displaying the scientific signifier as a regional paradigm of the signifier-in-itself: this is Plato's relation to Eudoxus, Leibniz's relation to Leibniz, Kant's relation to Newton, Husserl's relation to Bolzano and Frege, and perhaps Lacan's relation to mathematical logic."[35] The specificity of the philosophical gesture, vis-à-vis science, which is its most intimate other, would consist, then, in designating its signifiers (those of science) as belonging to a specific region of ideality—but only to one such region—which would call for its replacement in and by a philosophical ideality. That is the expression of its lack, which only philosophy would be able to suture. A scientific idea cannot be characterized as an "idea" except by means of a philosophical determination of that ideality. It would be philosophy's task alone to define the idea, thus, the traditional autopositioning of philosophy as region of all regions, arch-region, arch-discourse. In this manner, philosophy assigns lack to science, denying that it makes a clean cut in order to renew science's dependence on philosophy (which is certainly clear in the Platonic figures of the divided line and the cave).

Philosophy, Badiou continues, thus constructs a properly philosophical concept of science, but the problem is that science "cannot receive this mark."[36] Further, Badiou writes "that which, in philosophy, declares itself science, is invariably the lack of science. That which philosophy lacks, and that to which it is sutured, is its very object (science), which is nevertheless marked within the former by the place it will never come to occupy. We can claim in all rigour that science is the subject of philosophy, and this precisely because there is no subject of science."[37] In this way, philosophy alienates itself in lending its own subject to science.

In the language that I have chosen to adopt here, which is that of the response, we can thus say that philosophy is that discourse which attempts to respond for science, which is to say, that attempts to be ontologically responsible for it. It is a discourse that undertakes to make science respond by constituting itself as a placeholder for a response that science does not give. Hegel says it well, in the preface to the *Phenomenology of Spirit*, that mathematics is a dead construction, to the extent that it cannot reflect itself, and is, in a certain sense, mute. "For what is lifeless, since it does not move of itself, does not get as far as the distinctions of essence, as far as essential opposition or inequality, and therefore does not make the transition of one opposite into its opposite, does not attain to qualitative, immanent motion or *self*-movement."[38] Only a speculative grasp of mathematics, then, can reanimate it, make it speak. "Philosophy," continues Hegel, is on the contrary "that which posits itself and is alive within itself."[39]

* * *

Such a philosophical response, for Badiou, is necessarily challengeable because it is ideological. It coincides with the old idea according to which science cannot think, cannot stand for itself, posit itself, and appears to be for that reason the servant of philosophy. I do not plan to critique Badiou's argument here, nor to show how he has changed his position and point of view in his later works. I choose instead to end on the strange resonance that I have detected between his remarks and the admirable analysis of the question of a response made by Derrida in "Passions: An Oblique Offering."

* * *

Derrida asks himself what it means to respond, playing at the outset with his situation as an invitee of a colloquium asking what it means to respond to an invitation. Quickly, the question grows and gets deeper. It is a question of seeing how responding always leaves open the possibility of not responding, and how the "not responding," which is what concerns me here, cannot be interpreted in terms of lack. Not responding, writes Derrida, "would have nothing to do with a shortcoming, a lapse in logical or demonstrative rigor, quite the contrary."[40]

The nonresponse would be inscribed in every response as its characteristic secret, not as a lack. If it is in principle entirely possible to detect the nonresponse in the philosophical response itself, it is nonetheless necessary to leave the realm of philosophy in order to demonstrate it. It is thus not to philosophy that Derrida turns, nor is it, incidentally, to science, but to literature. If it is "impossible to respond," says Derrida, such an impossibility is not of the order of knowledge. But philosophy is linked to knowledge. Nor can that impossibility be of the order of science. That which does not respond, either in philosophy or in science, is not the unconscious, either. The nonresponse is what Derrida calls the secret. Once again, this secret belongs neither to psychoanalysis, nor to philosophy, nor to science.

The space of the secret, of the nonresponse, is literature. "Not that I want to reduce everything to it [literature]," says Derrida, "especially not philosophy." He continues, "But if, without liking literature in general and for its own sake, I like something *about it*, which above all cannot be reduced to some aesthetic quality, to some source of formal pleasure [*jouissance*], this would be *in place of the secret*. In place of an absolute secret."[41]

By literature, one must not understand "fine literature" or "poetry," but that "modern invention" that has mixed itself up with the "right to say everything. Literature thus ties its destiny to a certain non-censure, to the space of democratic freedom."[42] However, Derrida continues, "This authorization to say everything paradoxically makes the author an author who is not responsible to anyone, not even to himself. . . . This authorization to say everything acknowledges a right to absolute nonresponse."[43] Later, Derrida continues, "There is, in the exemplary secret of literature, a chance of saying everything without touching upon the secret."[44]

It seems to me that at a certain moment, two extremes meet. That the foreclosure, the absolute closure in which Badiou encloses the scientific signifier, can immediately invert itself into the absolute openness in which Derrida situates literature. Beginning with an enclosed whole or with an open whole, the signifier, in either case, does not respond. That is its secret—a secret that is not something hidden, which would be possible to detect, a secret that, maybe, does not even exist, or is nothing other than what it is: a nonresponse without lack. "There is something secret. But it does not conceal itself. Heterogeneous to the hidden, to the obscure, to the nocturnal, to the invisible, to what can be dissimulated and indeed to what is nonmanifest in general, it cannot be unveiled. It remains inviolable even when one thinks one has revealed it."[45]

There would then be, at the heart of the process of truth, a nonresponse, an indifference to response, a special space, separated but without suture, in which psychoanalysis and philosophy would certainly have their place, like all discourses, but which could not be subjected to them, precisely because that space, immensely open and immensely closed, is absolutely alien to the subject. The secret is "not of consciousness, nor of the subject, nor of Dasein."[46] Nobody's secret. Anonymity as secrecy.

* * *

For a long time, Althusserian Marxist discourse defined truth as a "process without subject." Badiou's essay "Mark and Lack" undeniably belongs to that history. Even if the times have changed, later premises "without subject" are still operating. Derrida's position, and more recently that of Quentin Meillassoux in his work, *After Finitude*, put forward the idea of an *absolute* nonresponse of the real, which is another way of saying the critique or deposing of the subject.[47] Even if their perspectives are very different, even sometimes opposed to each other, it is a matter of, in each case, circumscribing the space of a void, which is neither a lack nor a flaw, but that of science (Badiou), of the secret (Derrida), or of the real (Meillassoux). In *After Finitude*, it is the real that is said to not respond, in the form of those fossils or "arche-fossils" that do without us,[48] of that being of a noncorrelated world, indifferent to human presence, indifferent to any subject. "I will call ancestral, Meillassoux writes, any reality anterior to the emergence of the human species—or even anterior to very recognized form of life on earth."[49]

This deserted, neutral, dispossessed world is indifferent to the fact of being thought. Such a world is for Meillassoux the "absolute," radically separated—absolutus—from us, a world whose separateness from thought is such that it presents itself to us as nonrelative and "hence as capable of existing whether we exist or not."[50]

In regard of this indifference, traditional philosophy seems like a forced response, a determination to answer for, to make respond, or to respond in the place of the nonresponsiveness of the real. If we want to break with correlationism, that is with the subject-object relationship, with the forced response, we have to drown the philosophical subject, there also, in the secret flows and codes of mathematics and literature.[51]

Must we and can we today content ourselves with such visions of philosophy? A "psychoanalytic" vision first of all: Miller would say that the "secret" or the "nonresponse," the "nonlack" or the "noncorrelated," is precisely the space of the subject, a space that philosophy cannot nevertheless explore as it constitutes its unconscious, the unconscious of all discourses in general, including the logical one. Second, a counterideological epistemological vision: in fantasizing science's lack of subject, in exploring the cracks of its suture, philosophy just accomplishes its own subjection to sciences. Third, a deconstructionist literary vision: philosophy cannot unveil the secret of literature as it is precisely comprised in it. And more recently, the contemporary vision of the wall of the "real," against which the philosophical subject shatters and blows up.

What if the philosophical subject, contrarily to what one might think, was precisely unconcerned, indifferent to its own lack, to its zero ontological consistency, to its own emptiness? What if this indifference precisely constituted the "essence" of subjectivity, that is its "reality"? Commenting on Meillassoux's notion of ancestrality and "arche-fossil," Žižek rightly declares, "The true fossil is the subject itself in its impossible objectal status. The fossil is myself, the way the terrified cat sees me when it looks at me. This is what truly escapes correlation, not the In-itself or the object, but the subject as object."[52]

The individual subject is always excluded from the real subject. The subject's secret pertains to its essential objectivity, or thinghood, that is its own impossibility. It is true that philosophy explores a specific void, which is the blank square of subjectivity, but this void is meaningless, indifferent, it is what resists the real in the real. Detached from it. Unbound and inassimilable.

I have tried, in all my work, to situate this resistance. In *The Future of Hegel*, I analyzed the dialectical tension between subject and substance, and the way in which Hegel shows that subjectivity is always initially captured by its own inorganic neutrality. In *What Should We Do with Our Brain?*, I interrogated the cerebral subject, in its organic (non)responsiveness. Later, in The *New Wounded* and in *Ontology of the Accident*, indifference and coolness became the central actors of "subjective" life. More recently, my analysis of the epigenetic nature of the Kantian subject was an attempt at conceptualizing the materialization, thus the desubjectivation, of the transcendental.[53] Hegel had its own way of saying that philosophy would never be the subject of science: "Hitherto philosophy had not found its method; it looked with envy at the systematic edifice of mathematics, and, as we have said, borrowed it from it or helped itself with the method of sciences which are only an admixture of given materials, propositions of experience and thoughts—or it even resorted to the crude rejection of all method."[54] This statement remains true. It is also true that philosophy does not have the right to say everything. This is no secret, and this is fortunate. Philosophy to come has to invent the discourse of responsiveness and responsibility for its own unconcerned subject.

NOTES

1. This is a translation of an essay published in French. Catherine Malabou, "La science est-elle le sujet de la philosophie? Miller, badiou et Derrda se répondent," in *Rencontre autour de Catherine Malabou*, ed. Michaël Crévoisier (Besançon: Presses Universitaires de Franche-Comté 2017), 21–33.

2. Ten issues of *Cahiers pour l'Analyse* were published in France between 1966 and 1969. All of the essays can be found in a bilingual French-English edition on the website created by CRMEP at the University of Kingston, titled *Concept and Form: The Cahiers pour l'Analyse and Contemporary French Thought*, with indications of the original pagination. PDFs of the original version of each article are also available on that site. In this chapter, I refer to that site. Alain Badiou's essay "Marque et manque, à propos du zéro" appeared in *Cahiers pour l'analyse* 10, "La Formalisation" (Winter 1969): 150–73.

3. Jacques Alain Miller, "La Suture: Elements d'une logique du significant," *Le Cahiers pour l'Analyse* 1, "La Vérité," article 3 (February 1966): 37–49.

4. Jacques Derrida, *On the Name*, ed. Thomas Dutoit, trans. David Wood et al. (Palo Alto, CA: Stanford University Press, 1995).

5. Gottlob Frege, *Foundations of Arithmetic*, trans. J. L. Austin (Evanston, IL: Northwestern University Press, 1980).

6. Ibid., 85.

7. Miller, "La Suture," 42.

8. Ibid.
9. Ibid., 43.
10. Ibid.
11. Ibid., 44.
12. From there, he can deduce the other numbers by positing the proposition: "There is a concept *F* and an object *x* that falls under it, such that the number that corresponds to the concept *F* is *n* and the number that corresponds to the concept 'falls under *F* but is not equal to *x* is *m*.'" Which means the same as "*n* follows in the series of natural numbers directly after *m*." Frege, *Foundations of Arithmetic*, § 76. 89.
13. Miller, "La Suture," 45.
14. Ibid., 39.
15. Ibid.
16. Ibid., 47.
17. "The signifier [is] that which represents the subject for another signifier." Ibid., 48.
18. Ibid., 49.
19. Ibid., 39.
20. Ibid.
21. Ibid., 40.
22. Ibid., 46.
23. Ibid., 47.
24. Badiou. "Mark and Lack," op. cit., 161.
25. Ibid., 151.
26. The cut, Badiou says, takes place in language. By the three mechanisms of "concatenation," "formation," and "derivation," logic establishes the "series," which is to say the chain. See Badiou, "Mark and Lack," 151.
27. Ibid., 155.
28. Ibid., 156.
29. Ibid.
30. Ibid., 157.
31. Ibid.
32. Ibid., 160.
33. Ibid., 161.
34. Ibid., 163.
35. Ibid.
36. Ibid.
37. Ibid.
38. G. W. F. Hegel, *Phenomenology of Spirit*, trans. A. V. Miller (Oxford: Oxford University Press, 1977), 26.
39. Ibid., 27.
40. Derrida, "Passion," 11.
41. Ibid., 28.
42. Ibid.
43. Ibid., 28–29.
44. Ibid., 29.

45. Ibid., 26. In *The Postcard*, Derrida already affirms that lack is a metaphysical term, which has no meaning except in reference to presence. "The difference which interests me here," he says, "is that—a formula to be understood as one will—the lack does not have its place in dissemination." Jacques Derrida, *The Postcard*, trans. Alan Bass (Chicago: University of Chicago Press, 1987).

46. Derrida, *On the Name*, 30.

47. Quentin Meillassoux, *After Finitude: An Essay on the Necessity of Contingency*, trans. Ray Brassier (London: Continuum, 2009).

48. *Cf.*, ibid., 28.

49. Ibid., 10.

50. Ibid., 28.

51. Meillassoux's second book, *The Number and the Siren: A Decipherment of Mallarmé's Coup de Dé*, trans. Robin Mackay (Falmouth, UK: Urbanomic; New York: Sequence Press, 2012), is devoted to Mallarmé. After the developments on Cantor's transfinite in *After Finitude* came the deciphering of the secret code in Mallarmé's "Un coup de dé"—a secret code with nothing to reveal. Mathematics and poetry are circling around philosophy, desubjectivizing it, and upending its power of response.

52. Slavoj Žižek, *Less Than Nothing, Hegel and the Shadow of Dialectical Materialism* (London: Verso, 2012), 64.

53. I am referring here to some of my books: *The Future of Hegel: Plasticity, Temporality and Dialectic*, trans. Lisbeth During (New York: Routledge, 2005); *What Should We Do with Our Brain?*, trans. Sebastian Rand (New York: Fordham University Press, 2005); *The New Wounded: From Neurosis to Brain Damage*, trans. Steven Miller (New York: Fordham University Press, 2012); *Ontology of the Accident: An Essay on Destructive Plasticity*, trans. Carolyn Shread (London: Polity Press, 2012); *Before Tomorrow, Epigenesis and Rationality*, trans. Carolyn Shread (London: Polity Press, 2016).

54. G. W. F. Hegel, *Science of Logic*, trans. and ed. G. Giovanni (Cambridge: Cambridge University Press, 2010), 32–33.

BIBLIOGRAPHY

Badiou, Alain. "Marque et manque, à propos du zero." *Le Cahiers pour l'Analyse* 10, "La Formalisation," article 8 (Winter 1969): 150–73. http://cahiers.kingston.ac.uk/pdf/cpa10.8.badiou.pdf.

Derrida, Jacques. *On the Name*. Edited by Thomas Dutoit. Translated by David Wood et al. Palo Alto, CA: Stanford University Press, 1995.

———. *The Postcard*. Translated by Alan Bass. Chicago: University of Chicago Press, 1987.

Frege, Gottlob. *Foundations of Arithmetic*. Translated by J. L. Austin. Evanston, IL: Northwestern University Press, 1980.

Hegel, G. W. F. *Phenomenology of Spirit*. Translated by A. V. Miller. Oxford: Oxford University Press, 1977.

———. *Science of Logic*. Translated and edited by George di Giovanni. Cambridge: Cambridge University Press, 2010.

Malabou, Catherine. *Before Tomorrow, Epigenesis and Rationality*. Translated by Carolyn Shread. London: Polity Press, 2016.

———. *Ontology of the Accident: An Essay on Destructive Plasticity*. Translated by Carolyn Shread. London: Polity Press, 2012.

———. *The Future of Hegel: Plasticity, Temporality and Dialectic*. Translated by Lisbeth During. New York: Routledge, 2005.

———. *The New Wounded: From Neurosis to Brain Damage*. Translated by Steven Miller. New York: Fordham University Press, 2012.

———. *What Should We Do with Our Brain?* Translated by Sebastian Rand. New York: Fordham University Press, 2005.

Meillassoux, Quentin. *After Finitude: An Essay on the Necessity of Contingency*. Translated by Ray Brassier. London: Continuum, 2009.

———. *The Number and the Siren: A Decipherment of Mallarmé's Coup de Dé*. Translated by Robin Mackay. Falmouth, UK: Urbanomic; New York: Sequence Press, 2012.

Miller, Jacques Alain. "La Suture: Elements d'une logique du significant." *Le Cahiers pour l'Analyse* 1, "La Vérité," article 3 (February 1966): 37–49. http://cahiers.kingston.ac.uk/pdf/cpa1.3.miller.pdf.

Žižek, Slavoj. *Less Than Nothing*: *Hegel and the Shadow of Dialectical Materialism*. London: Verso, 2012.

Chapter 2

"The Plasticity of Writing"

Malabou on the Limits of Grammatology

Deborah Goldgaber

Catherine Malabou's work challenges philosophical habits of bracketing the body's materiality in accounts of subjectivity and lived experience. *The New Wounded* (2012) and *Self and Emotional Life* (2013) encourage us to consider how neuronal matter might be *internal* to accounts of subject formation and power while *What Should We Do with Our Brain?* (2008) asks how the "discovery" of brain plasticity should alter conceptions of practical and political limits. The title can also be read as a pointed question to a philosophical tradition that has tended to treat claims about the nature of matter either with disinterest—as too reductive to be of interest—or with skepticism—as if the materialist gesture amounted to reasserting the priority of *deterministic* matter over the *plasticity* of culture.

Malabou is one of the most influential "new materialist" Continental philosophers to argue that poststructuralist and deconstructive philosophies have thought materiality one-sidedly—foreclosing questions about the material world beyond the scope of cultural activity. While poststructuralist approaches have, broadly speaking, insisted on the *materiality* of language, cultural processes, and power—developing accounts of the corporeal effects of cultural processes—they have missed, Karen Barad writes, the *materiality* of matter. Even as theories of the body's discursive construction bring "matter and meaning into closer proximity" by challenging the view that (linguistic) representation and (material) reality are ontologically divorced, matter only seems to *matter* inasmuch as it is related to questions of signification or human activity.[1]

In *Plasticity at the Dusk of Writing* (2009) and *The End of Writing? Grammatology and Plasticity* (2010), Malabou argues that deconstructive philosophy's conceptual expansion of *texts* and *writing* was, from the perspective, of its materialist aspirations, "programmed to fail." Capable of powerfully underscoring the *materiality* of language, Derrida's generalization of writing, she argues, is simply incapable of thinking the materiality of *matter*. Malabou's critique of Derridean deconstruction offers a detailed diagnosis of the problem that Barad identifies, arguing that the dominance of the metaphor of inscription and writing—the "graphism" of poststructuralist thought—are impediments to the sort of materialist reflection Barad envisions. Where we think the body's matter *discursively* or *textually*, "matter" is necessarily limited to designating that which serves as the material substrate for cultural inscription. Whatever philosophical productivity these inscriptive models may have had, today they are exhausted; we need new concepts, new theoretical "motor schemas" that transcend the limitations of *text*, *code, discourse*, and *writing*.[2]

To grasp the full significance of Malabou's critique, it is crucial to recall the materialist promise of Derrida's *Of Grammatology* (*OG*). Grammatology should have been the name for a novel form of philosophical materialism, yoking the materiality of language together with the *textuality* of the material. Under the heading of grammatology, Derrida announced the *absolute* generality of "writing," where the latter was understood in an enlarged sense, distinct from empirical writing.[3] As absolutely general, *writing* would cover not only systems of human signification but also physical and biological systems, extending to the logic of life itself. Yet, as Claire Colebrook remarks, while deconstruction has given us some sense of what it means to talk about the *materiality* of the sign, it has not given an equally good sense for how to talk about the "textuality" of matter, nor a good sense for how to go from talk of one to talk of the other.[4]

Colebrook notes that "recent defenders of deconstruction" have returned to Derrida's generality claim, attempting to cash in its materialist import by introducing various nongrammatological supplements to Derrida's claims in *OG*. As an example of this strategy, Colebrook cites Cary Wolfe's interpretation of Derridean *différance* as

> compatible with a sense of life as self-differentiating, as nothing other than a dispersed system. [For Wolfe,] Derrida is not only compatible with systems theorists such as Luhmann and Maturana and Varela; he insists that the Derridean version of considering the world as a system of relations rather than original presence is required if systems theory is not to fall into one more Cartesian dualism.[5]

Here Wolfe supplements grammatology with a systems-theoretic approach in order to bridge the gap he identifies between the *différance* Derrida locates in linguistic systems—or texts narrowly construed—and nonlinguistic systems. This supplementary, "structuralist" approach, however, underscores the methodological weakness of the grammatological project that Malabou identifies. The principle or justification for *writing's* generalization was supposed to be contained in the production of a deconstructive concept of writing, *arche-writing. Arche-writing*, in turn, was supposed to pertain to everything traditionally thought as "outside the text"—to life and world at large. Yet, as Wolfe's approach demonstrates, this extension of writing to the extratextual has always appeared without adequate warrant, in need of supplementation. The principle of writing's *generality* must be secured by the truth or the generality of the proposed supplementary approach. Derridean writing will turn out to be generalizable to material systems or processes to the extent that (a) such systems or processes are explicable in terms of the target model (b) and that target model is itself explicable in terms of the "systems of relations" or modes of nonpresence writing designates.

According to Malabou, Derrida's *grammatological* strategy for thinking *différance* produces the problem theorists like Wolfe must then attempt to solve. However, much Derridean *writing* enlarges or expands the sense of writing—to get us to something like a "semiology without signs," or, the idea of a system of relations without original presence. This conceptual expansion retraces the inscriptive *form* of "narrow" colloquial writing.[6] It is impossible to think a form of general writing without reinscribing the most traditional of metaphysical distinctions: form/matter.

OG's generalization of writing, on Malabou's reconstruction, takes as its starting point the image of empirical writing, and it continues to rely on this image for its intuitive sense and confirmation, however far the concept goes away from the narrow, empirical image. To borrow from Malabou's own Kantian language, Derrida's conceptual modification of writing allows us to schematize empirical intuitions—but, these schemas will necessarily repeat the dualistic opposition inherent in the base image: that between graphic *form* and nongraphic *matter*. Grammatology, despite its declared generality, is constrained to think *writing* and therefore *différance* in terms of "frayage"—the marking-up, breaking-up, or breaching of a surface. "For Derrida one of the proofs that arche-writing remains essentially dependent on an understanding of the trace as a graphic inscription is the tenacity of the motif of frayage. Ache-writing must be understood (and can only be understood) in terms of frayage."[7] Yet, if *writing* everywhere and always implies material inscription, how can the matter of inscription be described in terms of this same *writing*? The grammatological interpretation of *différance* is incompatible with a genuine materialism. The project of deconstructive materialism today requires rejecting deconstruction's grammatological turn. Malabou suggests that neuroplasticity or "neuronal morphemes" and not empirical writing or code could serve as deconstructive materialism's new philosophical "motor-scheme."

In this chapter, I will question Malabou's claim that grammatology is incorrigibly "graphic." Not only can arche-writing be understand otherwise than in terms of frayage, but Derrida explicitly argues that it must. The most general concept of writing, he insists, covers both the popular concept of writing, which he refers to as graphic, but also nongraphic writing. "Arche-writing would be at work not only in the form and substance of graphic expression but also in those of nongraphic expression."[8] Indeed, Malabou's account of generalized writing as a "motor scheme" mischaracterizes the nature of Derridean *writing* precisely to the extent that the former but not the latter rely on what Derrida calls, in *OG* "a popular concept of writing." "The idea of a 'form of expression' linked by correlation to the graphic 'substance of expression' remains," Derrida insists, "very derivative with regard to the arche-writing of which I speak"[9] By contrast with this popular notion of writing, Derridean *arche-writing*, we shall see, requires thinking a notion of writing or inscription beyond the form/matter distinction, where material "inscription" is always already a modification of form, and every modification of form also an "inscription." Derridean *arche-writing* and Malabouian plasticity, I conclude, are much closer than suspected.

The plan for the rest of this chapter is as follows: I will first reconstruct Malabou's diagnosis of the failure of generalized writing to make good on its materialist promises, specifying why she believes that *plasticity* may succeed where *writing* has failed. Then, I will argue against Malabou's claim that Derridean writing is *graphic* in the sense she identifies. Renewed attention to Derrida's account of the "instituted trace" in *OG* will demonstrate a decisive break with any image of empirical, or *graphic* writing. Derridean *writing* was never graphic, in Malabou's sense, however much it leans on the empirical image of inscription for its development. The grammatological transformation of writing must involve the parallel transformation of our notion of inscription; to think inscription differently requires thinking the *plasticity* of the written trace.

GRAPHISM: MALABOU ON THE FAILURE OF GENERALIZED WRITING

Malabou's work on *plasticity* is less a break with deconstructive thought than a break with its *grammatological* limitations—limitations that she sees as delaying the formulation of a deconstructive materialism. Derridean deconstruction comes closest to articulating a philosophical materialism in its announcement of *generalized* writing in *OG*. Yet, for all *writing's* promise to bring matter and language into closer proximity, Malabou underscores, *OG* fails in its aims. This accounts for the fact that *OG* remains something of a philosophical orphan.[10] Few texts, at least until recently, have followed up on its suggestion to think matter—organic or inorganic—in grammatological terms, that is, in terms of *writing*.

While Derrida's interpreters have long emphasized the pertinence of *writing* to questions of language and signification—more or less bracketing the question of writing's "absolute generality"—readers trying to make sense of the generality claim have often adopted a transcendental interpretive strategy.[11] *Writing* is general insofar as it refers to the general conditions of *experience*.[12] As such, *writing* would pertain to every possible object of experience. The main problem with this strategy is Derrida explicitly disavows it. *Writing's* generality, Derrida insists, cannot be established by identifying it with the (transcendental) conditions that make experience possible. Of course, if *writing* were absolutely general, then *a fortiori* it would explain human consciousness, but—and this is crucial—the reverse is not the case. Derrida writes,

> [consciousness] is an emergence that makes the grammè appear as such (that is to say according to a new structure of non-presence) and undoubtedly makes possible the emergence of the systems of writing in the narrow sense. From "genetic inscription" and the "short programmatic chains" regulating the behaviour of the amoeba or the annelid up to the passage beyond alphabetic writing to the orders of the logos and of a certain homo sapiens, the possibility of the gramme structures the movement of its history according to rigorously original levels, types, and rhythms. But one cannot think them without the most general concept of the gramme.[13]

Here, Derrida is clear that consciousness, genetic inscription, alphabetic writing—*all* are to be explained in terms of *writing*, and not the reverse.

"Materialism," Malabou writes, "names the non-transcendental quality of form in general."[14] From this and other passages of *OG*, *writing* appears to be, for Derrida, a materialist principle, in the sense that Malabou has given this term. *Writing* is not a form imposed on the world to explain its appearance or intelligibility; in the passage above, it appears as a kind of immanent history of life and its forms or "morphemes" ("the possibility of the gramme structures the movement of its history").[15] The gramme is, by all appearances, self-structuring, producing itself differently, "according to rigorously original levels, types and rhythms."[16] Consciousness and genes, annelids, and alphabets are distinctive levels of the emergence of the gramme. Accounts of these phenomena must respect the differences of levels and types, but they all require "the most general concept of the *gramme*."[17] The latter will necessarily exceed any of its "historical" instantiations. Consciousness is but one determined instance of the *gramme's* emergence—"undoubtedly" one that makes the structure of the *gramme* appear (for us)—but transcendental reflection on the structure of consciousness will not get us to the most general concept of the *gramme*. As Derrida shows in *OG*, reflection on empirical phenomena is also required; graphic writing is among the privileged points of departure.

If the *gramme* or *arche-writing* was to provide the sort of explanatory power Derrida asserted for it, what accounts, Malabou asks, for grammatology's failure—that is, its failure as novel philosophical materialism? Grammatology, as the science of general writing, *should* have linked the general conception of the *gramme* with innumerable empirical phenomena beyond consciousness, language, and signification. If the *gramme* covers, as Derrida writes, the *pro-gramme* of life, then it should have been possible to conceive biological life in grammatological terms. If such grammatological inquiries never saw the light of day, how should we account for this lack of uptake? Malabou is right to insist that this failure must be explained or we will have little sense of how to revise the approach for a deconstructive materialism today.

What accounts, then, for the failure of general *writing*? To answer this question Malabou first dispenses with what she argues is a red herring of sorts. Readers of *OG* will recall that Derrida was the first to have announced grammatology's failure. Grammatology, he insisted, could never be a positive science. However, as Malabou underlines, Derrida's reasons for why grammatology could never be a positive science do not bear on his claims about the absolute generality of *writing*—nor on the philosophical status or productivity of the concept of the *gramme*. If anything, the opposite is true: the less grammatology looks like the name for a positive science (like linguistics, or biology), the more it looks like the name for a distinctive form of philosophical materialism. If grammatology is programmed to fail—as Malabou insists it is—it is not for the reasons that Derrida gives. But it is also not—or so I shall argue—for the reasons that Malabou gives.

Malabou argues that the generalization of writing involves the expansion of the narrow image of writing—it involves the production of writing as a "motor scheme." Indeed, the first chapter of *OG*, "The End of the Book and the Beginning of Writing," reports on the expansion or inflation of writing that was, at the time of its writing (1967), in the "air du temps." Derrida notes,

> *We* say "writing" for all that gives rise to an inscription in general, whether it is literal or not and even if what it distributes in space is alien to the order of the voice: cinematography, choreography, of course, but also pictorial, musical, sculptural "writing." . . . [Writing] describe(s) not only the system of notation secondarily connected with these activities but the essence and the content of these activities themselves. It is also in this sense that the contemporary biologist speaks of writing and pro-gram in relation to the most elementary processes

> of information within the living cell. And, finally, whether it has essential limits or not, the entire field covered by the cybernetic program will be the field of writing.[18]

Already, without the least "deconstructive" intervention, the semantic field of "writing" seems enormous—and unwieldy. Writing covers all that could give rise to a mark—all that is structured like a mark—information in its broadest sense. In this context, we should note that Derrida's reference to cybernetics is more than casual.

In the third chapter, "Grammatology as a Positive Science," Derrida returns to the proposed postwar science whose unity would be founded on the generality of code, or the "program."[19] *Grammatology* can be read as offering the philosophical foundation for the cybernetic project—its statement of method—grounding the latter's generalist aspirations by laying out the conditions for such a science. "Grammatology" would be the name for a critical or well-founded cybernetics and *arche-writing* would be its core concept. Derrida, in turn, would be (or should have been) the Saussure of cybernetics by giving the latter its conceptual object, or "sign," even as cybernetics would come to enfold all of semiotics, linguistics, and other derivative disciplines.

However, Derrida argues, the condition for establishing the science of generalized writing turns out to be the condition for its failure. Why does *arche-writing* undermine the condition of a positive science? The most general concept of the gramme covers a field without obvious limit, exceeding the purview of any science. If Derrida is right and science must have a definable object or field, writing's excessive generality would settle the question of the possibility of grammatology *as a positive science*. But this "failure" leaves open and even supports the possibility of another sort of success: pursuing grammatology as a novel *philosophical* or metaphysical project—grammatology as deconstructive materialism.

In her essay "Grammatology and Plasticity," Malabou suggests that grammatology should have been the name for such a philosophical project but that such a project could never be founded on a grammatological foundation. The philosophically salient issue today is not whether grammatology can be a science, but whether a notion of generalized writing could help us to think through the nature of material and organic processes. While Derrida intended for grammatology to yield the most general concept of *différance*—demonstrating its pertinence beyond the confines of linguistic texts—it fails, on her diagnosis, because of *writing's* inherent limitations as a motor scheme.

What is a "motor scheme" and how does a motor scheme pertain to Derrida's notion of generalized *writing*? Malabou defines a motor scheme as a certain articulation of the empirical and conceptual, involving inflation or generalization.[20] If, as Derrida notes, empirical writing or notation refer narrowly to the spatial distribution of linguistic marks, writing *qua* motor scheme, had expanded (its reference) to include the genetic code—the "program of life"—and to the whole field of cybernetics, or what we might today call information technology. The image of empirical writing "pumps our intuitions"—to borrow a phrase from Daniel Dennett—helping us to make explicit features, relations, and properties that may be useful to thinking a whole class of phenomena.[21] But if a motor scheme incites analogical thinking, the intuitions it makes possible are inherently limited; "writing's" schematic power depends upon the isomorphism of the relata. If Derridean *writing* involved constructing a philosophical concept that would makes *explicit* the features, properties, or relations implicit in the image of empirical writing, then it would account for both the possibility of the inflation of empirical writing and the limits of this inflation. *Writing* would specify and regulate the possible transformations of the empirical image (to those that preserve the relations implicit in the empirical image) by making explicit the conditions for its "nonliteral" use. Construed as a motor scheme, even an epochal motor scheme, *writing* seems to foreclose rather than warrant its own generality.

As Malabou notes, if Derrida chose empirical writing as a privileged image for *différance* or the *trace*, this is due in no small part to the inflation of writing that was already in the cybernetic *air du temps*. The cybernetic expansion of writing had already made it possible, even mundane, to think of the real or original—that which could give rise to notation—as *original* writing (e.g., dance/movement as choreography)—that is, according to a certain system of nonpresence. For example, thought as *writing*, motion is not only something that could be captured—by any manner of what, today, we would call motion-capture technologies—but also something "always already" so captured or "spaced," unfolding within the calculable coordinates of a Cartesian space. However, as Malabou underscores, if *writing* warrants the thought that "all that which can give rise to an inscription" is always already inscribed—reduceable to *spatial* distribution of graphic marks—does this not require us to think of matter both as that which is marked-up or inscribed and that which makes possible or "spaces" this distribution? In other words, does *arche-writing* not designate matter in the most classical metaphysical terms, as passive receptacle for form? Does the image of empirical writing from which general writing has apparently gained so much distance not consign matter to the extratextual, while determining it as the transcendental condition for *writing*? By the same token, if the meaning of *différance* were limited to *arche-writing*, how could we think the matter in which writing is spaced in terms of *différance*? It is for this reason that Malabou insists, as we saw above, that *arche-writing* can only be thought as *frayage*.

The problem with *arche-writing* and the grammatological strategy more generally, according to Malabou, is the incorrigible *graphism* of writing. Empirical writing is an image of *différance*, to be sure, but it is also an image of the opposition between form and matter. This helps to explain why *arche-writing* launched so many projects aimed at thinking the materiality of writing, but—as we saw Colebrook noting above—almost no projects capable of thinking the *textuality* of matter. To touch matter, writing would require something like the notion of a *nongraphic writing*. But the latter, Malabou suggests, cannot be schematized. Despite Derrida's claims to the contrary, we cannot imagine a nongraphic writing any more than we can imagine a square circle. Or, as Malabou puts it, the "plasticity of writing" will not extend so far. What is required is a *nongraphic* notion of *différance*, a new motor scheme for thinking materiality as *différance*. Malabou, as is well known, proposes neuroplasticity.

To sum up Malabou's diagnosis of grammatology's failure: however much Derridean *writing* displaces and expands the narrow, colloquial notion of writing in order to account for processes beyond language (e.g., choreography, photography), *writing's* productivity as a materialist principle—its capacity to help us think of matter in terms of *différance*—will always founder on its empirical point of departure. *Writing* could produce novel conceptual schemas with which to rethink ideal and material processes only insofar as these processes could be read as *variations* on the graphic mark. Generalized *writing*—on Derrida's own terms—should have accounted equally for language and life, for substrate and inscription—yet it approaches its explanatory limits at the moment it attempts to think the relation of matter and matter.

PLASTICITY BEYOND GRAPHIC *WRITING*

Malabou does not claim that it is implausible to think of materiality in deconstructive terms. Indeed, she endorses the project of thinking materiality in terms of *différance* or the *trace*, that is, in terms of forms of nonidentity that would require us to rethink ontological categories of substance and bodies. Plasticity is, for Malabou, the thought of the nongraphic *trace*, allowing us to think materiality in way that is constrained by neither the formal properties peculiar to signifying elements nor to what Derrida called *forms* of presence. Heir to the deconstructive notion of writing, but capable of thematizing what *writing* has so far obscured, plasticity names the possibility of morphogenesis—the original modifiability of matter.

Malabou proposes *plasticity* as a motor scheme that makes explicit what is implicit in the empirical image of neuronal plasticity, the intrinsic power of matter to take on, lose, and give itself form. Neuronal plasticity gives us a nongraphic model through which to think the retention and appearance of difference as a modification of form. Moreover, as Malabou underlines, contemporary neuroscience no longer thinks neuronal modification—as did Derrida's Freud—in terms of breaching or *frayage*. The brain does not encode or retain information in the way that a clay tablet retains marks on its surface. Neurons connect by modifying their resistance. Neuronal matter is formed by and defined in terms of a *double* capacity for retention. Matter retains modification, by lowering or raising its resistance, on the one hand, and these differences (changes in resistance) themselves form chains, "linkages," or "assemblies":

> The model of reformation, of recomposition, substitutes itself for the model of frayage: "it becomes plausible that such assemblies, made up of oscillatory neurons with high spontaneous activity, could recombine among themselves." "Linkages," "relationships," "spider webs," such are the configurations that the networks of nerve information take. It appears then that the synaptic openings are definitely gaps, *but gaps that are susceptible to taking on form*. The example of neurobiology is only one example of the fecundity of plasticity in the real. *We could surely call together other examples that today show that traces take on form*.[22]

As we have just seen, writing cannot be absolutely generalized if it presupposes the distinction between inscribed *form* and material *substrate*. If the latter is the condition of any mark, it cannot be understood in terms of writing. Yet, it *is* possible to think materiality in terms of *différance*—Derrida's term for the spacing and temporalization of differences—if we think of nongraphic *différance*, or "traces [that] take on form."[23] Indeed, Malabou is perfectly right, for the very reasons she adduces, to argue that deconstructive materialism requires thinking the nongraphic trace. But, she is wrong, I will now argue, to think that Derridean *writing* is *graphic*. Grammatological *arche-writing* was never graphic, or so I shall demonstrate below. If Malabou designates the "plasticity of writing" as both the condition for the expansion of the narrow concept of writing and the inherent limits of this expansion (to instances of "frayage"), *arche-writing* implies an entirely sense for the phrase. In the latter case, the "plasticity of writing" refers to the essential plasticity or modifiability of written texts, their capacity to retain and be inscribed by other texts. Texts are themselves the matter or substrate "in" which other texts are inscribed. Texts are doubly inscribed—informed forms or patterned patterns—and hence indissociably form and matter. Or, as Derrida writes in *OG*,

> It should be recognized that it is in the specific zone of this imprint and this trace . . . that differences appear among the elements or rather produce them, make them emerge as such and constitute the *texts*, the chains, and the systems of traces. These chains and systems cannot be outlined except in the fabric of this trace or imprint.[24]

With the same sort of curious philosophical hindsight that allowed Malabou to find secreted in the texts of Hegel and Heidegger the explosive power of plasticity, we can find in Derrida a conceptual sleeper cell: the notion of the plastic trace. Drawing out the plasticity of the trace is, I argue, important for the future of deconstructive philosophy. It also works as an important supplement to Malabou's conception of plasticity, specifying a formal schema—one that I will describe as a retentive mechanism—in terms of which we can conceive material plasticity.

ARCHE-WRITING AS NONGRAPHIC *WRITING*

Derrida's generalization of writing, according to Malabou, works by gives us something like "a semiology without signs." Rather than showing that everything is "structured like a language"—a position sometimes attributed to Saussure—generalized writing seeks to show how everything is structured like *writing*—caught in the play of *iterability* or *différance*. To get from the linguistic sign to generalized writing, Derrida must refashion or reform the notion of a sign form. Malabou notes that grammatology must first "liberate semiology from linguistics," from its limitation by a concept of signification understood in terms of linguistic meaning. Grammatology would then be a semiology, no longer limited by language, in terms of which we could, presumably, think the nature of life from the "annelid" to the human.[25]

Malabou's diagnosis of graphism misses misreads the direction of Derrida's reading of Saussure in the second chapter of *OG*. It is quite true that in "Linguistics and Grammatology," Derrida readily acknowledges the formalist aspects of Saussure's account of the linguistic sign—not least of all, distilled in the latter's dictum that language is "a form and not a substance."[26] Yet, he finds much in Saussure's account that resists this formalism. Indeed, reexamining Derrida's reading of Saussure in the light of Malabou's critique, we find him remarking the tension in Saussure's account, between the formalist tendency, on the one hand, to abstract from the "substance of expression" and, on the other hand, what we can call a properly grammatological tendency, to "rigorously isolate the bond that links *the play of form* to the substance of graphic expression."[27] Indeed, as Derrida shows in his reading, writing often serves Saussure as the image for the *indissociability* of form and substance. It is for this reason that, in order to explain the form of the arbitrary sign, Saussure must make "a necessary passage through the imprint."[28]

To be sure, empirical writing and inscription provides an empirical image of this bond—but only to some extent. The image of writing misleads; it is only so plastic. The essential feature of an iterable mark is to be unbound from any particular spatiotemporal context—to function in contexts that cannot be governed by any one of its appearances. This essential repeatability of the written mark permits us to abstract from the constitutive materiality of its form. Thus, we have the tendency to think of a written text, now from the perspective of its (repeatable) form and (specific) materialization, without generating any intuitions about the "necessity" or "logic" that might bind "the graphie"—a possible unit of writing—to the graphic substance in which it is expressed.

What Derrida finds unusual, and productive, in Saussure's account of the arbitrary linguistic sign, is the latter's insistence on the indissociability of the two faces of the sign. The signifier and the signified do not preexist the sign, but are differences constituted by the bond—each face *like* an original imprint or inscription appearing in and through the other. As Derrida writes, "The meaning of différance in general would be more accessible to us if the unity of that double passage appeared more clearly."[29] Thus, Derrida turns to Saussure, not because the latter provides us with a congenial, formalist account of the sign, but quite the opposite, because he sketches an account of a novel notion of form, based on a double inscription or retention.

On Derrida's original reading of Saussure, the latter elaborates a notion of an essentially retentive form—a form in which "inscription" and "materialization" are absolutely solidary. That Saussure describes the sign as essentially retentive is not obvious. Indeed, it has not been obvious even to some of Derrida's best interpreters. For Saussure, the arbitrary, linguistic sign is the indissociable unity of signifier and signified. Derrida argues that Saussure's insight into to a form whose mark is an indissociable bond requires recourse to the thought of graphic writing or inscription. Crucially, this necessary "passage through the imprint" is not due to the fact that the mark must be inscribed, in sound or another material medium such as ink/paper. Such a purely formal conception of the sign "still operates with a popular concept of writing." By "popular," Derrida explicitly refers to the narrow, inscriptive designation of writing in contrast to which "the arche-writing of which I speak . . . would be at work not only in the form and substance of graphic expression but also in those of nongraphic expression. It would constitute not only the pattern uniting form to all substance, graphic or otherwise, but the movement of the sign-function."[30] "The pattern uniting form to all substance" cannot be conceived, that is, according to the "popular concept of writing"—namely, inscription. In contrast to the formalist interpretation, with its abstraction from the materiality of expression, Derrida proposes a reading of Saussure's arbitrary sign form (and its indissociable unity) as "instituted trace."

We have already gotten some sense for why Derrida substitutes the term "instituted trace" for Saussure's "arbitrary sign." The form in question is instituted—it is produced "arbitrarily" as opposed to given "naturally" or "motivated." It is a trace because it is composed of elements (e.g., "signifier" and "signified") that do not exist prior to or outside their "institution"—the elements appear only in and through their mutual inscription or retention. The bond is indissociable precisely because it is constitutive of what it bonds. Signifying elements appear in their original form as inscriptions or transcriptions. Both the signifying and signified elements are produced in being reciprocally inscribed. But inscribed in what exactly?

Phenomenologically, in speech, we hear contrasts rather than sounds. Sound as such does not matter in speech; the contrasts "in" sound matter. Signifiers are just these contrastive elements. This should not lead us to say that speech is like a writing because sound is the material "substrate" on which contrasts are written or inscribed. Rather speech is composed of signifying elements rather than sound, in the form that speech implies; "sound" is retained as a contrast. If Saussure distinguishes speech from writing it is because he believes that, in speech, the contrastive elements in speech are produced in and through their indissociable bond with a heterogeneous contrastive element. Writing, by contrast, represents the sign one-sidedly. Its empirical image does not give us access to the retentive mechanism that produces or constitutes contrasts. In general, then, the sign form produces or elaborates contrastive elements.

An illustrative analogy will help make the point. In the mundane traffic light, red and green are signifying elements only insofar as they are construed as contrasts rather than colors. On Saussure's view, the contrasts or signifying elements appear only when articulated by heterogeneous contrasts, in our case, the contrast of "stop" and "go." Red/stop and green/go form a coordinated signifying system, which can then be represented in a graphic system by any suitable contrastive sets (e.g., 1/2; #/$). Red and green can appear as signifying elements—rather than mere colors—only insofar as they stop being what they were, namely, "red" and "green" and become arbitrary, or take on a signifying value. But they become arbitrary or take on a value, only if they retain or are imprinted by a set of heterogeneous variations, in our case the difference between movement or rest. The latter, in turn, are articulated as contrasts, having a binary value, insofar as they are articulated by or retained "in" a heterogeneous set of contrasts.

Of course, this "analogy" does not quite get us to the question of where we might find the sort of "double inscription" or "instituted traces," which produce the sort of contrastive elements alluded to above. Quite fortuitously, Malabou in her account of neuronal assemblies as nongraphic traces offers us, I believe, a very suggestive reply. Quoting Valentino Braitenburg, she writes,

> "The 'things' and 'events' of our experience . . . do not correspond within the brain to individual neurons, but to groups of neurons called 'cell assemblies'."

> These groups may form in a single area of the brain, but they may also be formed through an overlap of different areas. The assemblies amount to " 'units of meaning' as they appear as 'morphemes' in linguistics." Neuronal assemblies would thus form a logic of signification, creating a "morphematic" distribution of the configurations at the root of our ideas." . . . [w]hat Braitenberg and so many others since Hebb call a "trace" is a modification of form that corresponds to a plastic coding of experience.[31]

Malabou, of course, takes this "plastic coding of experience" to be non-grammatological. Yet, this neuronal account of how differences appear and take on form, through their retention ("coding") in synaptic modifications, is precisely grammatological in Derrida's sense. "Even before it is linked to incision, en-graving, drawing, or the letter, to a signifier referring in general to a signifier signified by it, the concept of the graphie [unit of a possible graphic system] implies the framework of the instituted trace . . . The instituted trace cannot be thought without thinking the retention of difference."[32]

Indeed, it is this form—that of the instituted trace—that Derrida claims is generalizable—a novel way of thinking the retention of difference that can be extended, beyond the organization of semiological systems to the organization of life and its material substrates. The principle seems to be the following: any set of "morphometric" differences can retain heterogeneous differences only if they are modified by them in a way that preserves or transcribes their pattern. Color can function in and as a signifying system if one of its qualitative dimensions—hue, saturation, intensity—is selected to stand in for a heterogeneous set of differences. If phonetic writing can "represent" speech, this is because it has the resources or affordances to borrow a term from Gibson's ecological thinking and to reproduce in its very "textuality" the affordances (pattern of contrasts) in speech. We can, that is, unpack or retrieve speech from phonetic writing. For Derrida, the mark of a signifying system is to be understood in terms of this reciprocal retention of differences. But this mechanism of retention is not limited to signifying systems. Signifying systems like language are exemplary in a double sense: (1) they give us an empirical image of the trace as retentive mechanism and (2) they demonstrate the signification is at its root the retention of differences. What is primary or original, then, is not the production of signification; signification is the effect or, better, the possibility of a generalized mechanism of retention. "Without a trace retaining the other as other in the same, no difference would do its work and no meaning would appear."[33] Retention is not for the sake of meaning; retention, or the possibility of *arche-writing*, is primary. Derrida's claim with respect to the absolute generality of writing entails that such affordances go all the way down.

CONCLUSION: THE PLASTICITY OF INSCRIPTION

For Derrida, the reciprocal inscription of difference that Saussure describes passes through, modifies, and transforms traditional notions of form. Derrida selects "trace" to designate the properties that Saussure gives to the arbitrary sign as a mechanism of "original retention." The trace structure, he argues, allows us to account for the necessity that leads Saussure to speak of the sign form in terms of imprint: *différance* is therefore the formation of form. But it is on the other hand the being-imprinted of the imprint (Derrida 1997, 69).

In Saussure's account of the sign, its two faces are indissociably bonded; the signified and signifier are differences that appear to be reciprocally "inscribed." However, if the form Saussure describes is constituted by the "inscription" or "imprint" of one set of differences in other set of differences, such a mechanism of retention necessarily break with any model of empirical inscription (e.g., frayage, engraving). Indeed, as Derrida argues, the indissociable bond that Saussure describes is impossible to account for within a metaphysical framework fixed on the traditional distinction between form and matter—the very binary in which, Malabou claimed, "writing" remains caught. Derrida writes,

> it should be recognized that it is in the specific zone of this imprint and this trace . . . that differences appear among the elements or rather produce them, make them emerge as such and constitute the texts, the chains, and the systems of traces. These chains and systems cannot be outlined except in the fabric of this trace or imprint.[34]

The elements or forms in question are constituted in and through the "fabric" of what is always already a trace. The fabric (or text) is a system of traces that always already bears (as its form or structure) the imprint or the trace of other traces—and this "fabric" is (for Derrida) at once the appearance, the spatialization and temporalization, of differential elements and the production of these differential elements through a displacement or modification of what was always already "imprinted." "This last concept [that of the imprint] is thus absolutely and by rights 'anterior' to all physiological problematics concerning the nature of the engramme [the unit of engraving]."[35]

I have endeavored to show Derridean writing implies an expansion and modification of the category of inscription. However, as the foregoing suggested, the sort of inscription that Derrida aims to bring out has difficulty finding a place in the conceptual framework we use to describe both material objects and ideal objects. This ontology of substances, spatiotemporal particulars, and causal chains on the one hand, and ideality, temporality, and meaning on the other, cannot capture the way that writing and the trace do their work.

"Without a trace retaining the other as other in the same, no difference would do its work."[36] The form of inscription implied by the trace just is the retention of the other as "other in the same." Malabou is correct to argue that if a trace is thought as form materially inscribed or instantiated, then it is logically distinct from materiality. On the other hand, as I have just argued, nothing prevents us from thinking of the trace as materially "inscribed" in another trace—not iterated on a surface or some material, but retained in and through its (material) modification of another set of differences. Why not think the trace as informed forms, en-patterned patterns, as precisely the modifiability of form in its retention of heterogeneous differences? Would this not solve the interpretive problem Malabou diagnoses while giving us an account of the written trace that presupposes its essential plasticity (modifiability)—justifying the extension of writing that leads from one level of life's organization to another?

If we follow this suggestion, we can see that the concept of trace that seems to justify the claim of its absolute generality also implies the plasticity of the trace. Or, as Derrida showed in his reading of Saussure, modifiable traces (as in-formed texts) form chains or "texts" related to each other in terms of a "reciprocal conditioning." Just as easily we could substitute "texts" for "assemblies" of modifiable, iterable traces—language that, as we saw above, Malabou believes expresses the nongraphic plasticity of the trace. This capacity to modify and be modified gives us a nongraphic model through which to think the retention difference.

It was precisely the insistence that the trace was a repeatable form materially inscribed that blocked any possible conception of understanding materiality in terms of the trace. Malabou's reading of the limits of grammatology proves invaluable for making that assumption clear. But in her diagnosis she describes, arguably, not the limits of grammatology but the limits of prevalent interpretations of *arche-writing*.

NOTES

1. Karen Barad, *Meeting the Universe Halfway: Quantum Physics and the Entanglement of Matter and Meaning* (Durham, NC: Duke University Press, 2007), 64

2. Catherine Malabou, "The End of Writing? Grammatology and Plasticity," *The European Legacy* 12, no. 4 (2007): 433. Malabou introduces the concept of motor schemes to describe the philosophical productivity of certain concepts. "Code," "discourse," "text" represent such motor schemes, as does her own "plasticity." "Mechanism," "clocks," and "economy"—especially viewed from the point of view of its thermodynamic roots—represent other possible motor schemes. Malabou's notion of a motor scheme asks us to consider the philosophical productivity of certain images—their capacity to serve as what Daniel Dennett has called intuition pumps. Because for Malabou "the real" outstrips our concepts, motor schemes help us to generate new

concepts from old ones. It may be productive to think of motors schemes in terms of what Kant called, in the *Critique of Judgment*, aesthetic ideas. Kant distinguishes the cognitive role of the latter from the normal use of concepts. If normally the role of concepts is to bring intuitions under a determinate rule, the role of aesthetic idea is to generate intuitions. The latter "prompt much thought"—our cognitive apparatus revels in the associative indeterminacy of such ideas. This account of motor schemes is central to Malabou's interpretation of Derridean writing.

3. To mark this distinction, I will write Derridean, generalized *writing* with italics, reserving the nonitalicized "writing" for the narrow, empirical concept.

4. Claire Colebrooke, "Matter without Bodies," *Derrida Today* 4, no. 1 (2011): 2.

5. Ibid., 4.

6. Malabou, "The End of Writing?," 436.

7. Catherine Malabou, *Changing Difference: The Question of the Feminine in Philosophy* (Cambridge: Polity Press, 2011), 61.

8. Jacques Derrida, *Of Grammatology* (Baltimore: Johns Hopkins University Press, 1997), 60.

9. Ibid.

10. Malabou, "The End of Writing?," 432.

11. This interpretation is grounded in Rodolphe Gasché's classic *Tain of the Mirror* (1986). The latter text argues that terms like *différance* and *trace* refer to those structures that produce phenomenological experience but which speak against the "closure" of this experience in something like the "unity" of consciousness. On my reading, Gasché's interpretation—which explains the structure of lived experience in terms of *différance*—is perfectly compatible with Derrida's notion general writing, but because it limits différance to an account of consciousness it is incompatible with Derrida's generality claim. We cannot get from Gasché's account to a generalized *writing* without a more robust account of *how* writing is generalized. Readers such as Bernard Stiegler have argued that Gasche's reading is correct and that, despite Derrida's claims, we must read *writing* as limited to consciousness. See also Bernard Stiegler, *Technics and Time* (Palo Alto, CA: Stanford University Press, 1994).

12. This has led Speculative Realists to diagnose deconstruction as a correlationist philosophy, where correlationism refers to any philosophical position that articulates the conditions of our access to the world in terms of a correlation between world and consciousness. On the Speculative Realist reading—which I argue elsewhere is wrong—*writing* is functionally equivalent to philosophemes like intentionality and phenomenon. Or, following Gasché, *writing* is shorthand for the sort of transcendental activity that accounts for experience. If it differs from other philosophies of experience, it is only insofar as it privileges the constitutive role of language.

13. Derrida, *Grammatology*, 85.

14. Catherine Malabou, "Whither Materialism? Althusser/Darwin," in *Plastic Materialities: Politics, Legality, and Metamorphosis in the Work of Catherine Malabou*, ed. Brenda Bandhar and J. Goldberg-Hiller (Durham, NC: Duke University Press, 2015), 48.

15. Derrida, *Grammatology*, 85.

16. Ibid.

17. Ibid.

18. Ibid., 9.
19. Ibid., 84.
20. Malabou, "End of Writing?," 437.
21. See Daniel Dennett, *Intuition Pumps and Other Tools for Thinking* (New York: Norton, 2014).
22. Malabou, "End of Writing?," 440. Emphasis mine.
23. Ibid.
24. Derrida, *Grammatology*, 65.
25. Malabou, "End of Writing?," 432.
26. Derrida, *Grammatology*, 67.
27. Ibid., 59.
28. Ibid., 62.
29. Ibid.
30. Ibid., 60.
31. Catherine Malabou, *Changing Difference: The Feminine and the Question of Philosophy*, trans. Carolyn Shread (Cambridge: Polity Press, 2011), 59.
32. Derrida, *Grammatology*, 46.
33. Ibid., 62.
34. Derrida, *Grammatology*, 65
35. Ibid., 65.
36. Ibid., 62.

BIBLIOGRAPHY

Barad, Karen. *Meeting the Universe Halfway: Quantum Physics and the Entanglement of Matter and Meaning*. Durham, NC: Duke University Press, 2007.

Colebrook, Claire. "Matter without Bodies." *Derrida Today* 4, no. 1 (2007): 1–20.

Dennett, Daniel. *Intuition Pumps and Other Tools for Thinking*. New York: Norton, 2014.

Derrida, Jacques. *Of Grammatology*. Baltimore: Johns Hopkins University Press, 1997.

Gasché, Rodolphe. *The Tain of the Mirror: Derrida and the Philosophy of Reflection*. Cambridge, MA: Harvard University Press, 1986.

Malabou, Catherine. "The End of Writing? Grammatology and Plasticity." *The European Legacy* 12, no. 4 (2007): 431–41.

———. *Plasticity at the Dusk of Writing: Dialectic, Destruction, Deconstruction*. Translated by Carolyn Shread. New York: Columbia University Press, 2010.

———. *Changing Difference: The Question of the Feminine in Philosophy*. Translated by Carolyn Shread. Cambridge: Polity Press, 2011.

———. "Whither Materialism? Althusser/Darwin." In *Plastic Materialities: Politics, Legality, and Metamorphosis in the Work of Catherine Malabou*, edited by Brenda Bandhar and J. Goldberg-Hiller, 47–60. Durham, NC: Duke University Press, 2015.

Chapter 3

The "Image of Thought" at Dusk

Derridean-Husserlian Responsibility, Destructive Plasticity, and the Manifesto

John Nyman

One of the major interests of Catherine Malabou's thought is the reestablishment of form and visibility as operative philosophical concepts. For many readers this interest is apparent in Malabou's insistence on treating the neurobiological brain—an entity that is by definition both materially constituted and empirically accessible—as the privileged representation of thought and the mind. However, Malabou also takes form and visibility as central themes in texts that owe very little to empirical science, most notably *Plasticity at the Dusk of Writing*, which both summarizes and extends the argumentative positions Malabou develops throughout her earlier philosophical work. While remaining very far from empiricism or scientism, *Plasticity* forwards an emphatic, concentrated critique of the deconstructive philosophies of Jacques Derrida and Emmanuel Levinas on the basis of one relatively straightforward claim: neither of these philosophers, especially read according to their central concepts of writing and the trace of the other, is capable of accounting for the phenomenon of visible, material transformation. By refusing to address the fundamental reality of form and its transformability, deconstruction ignores the medium upon which its own inscription actually takes place. As Malabou puts it, "Perhaps the other makes 'its entry' as a line that scratches, tears, slices, and striates. But if I were not willing to receive it, if my ontological metabolism were not ready for it, I would have no reason to welcome the phenomenon. *One must be in good shape to welcome the trace. [Il faut être en forme pour accueillir la trace.]*"[1]

At the same time, Malabou does not merely dismiss the lessons and legacies of deconstruction. Far from characterizing Derrida's philosophy of writing as a simple misstep, Malabou insists on its historical validity as one of the foundational faces of a philosophical tradition that is itself characterized by metamorphosis. If Malabou's "is a question of recognizing [the] *visibility*" of Derrida's "mark" or "fold," which for her "are in fact grooves, types of relief,"[2] her contribution is itself premised on these marks having already emerged in the negative, invisible events of inscription thematized by Derrida. Malabou situates plasticity "at the dusk of writing" because it emerges only in conjunction with writing's decline. Moreover, this decline is not externally imposed, but inherent in the movement by which writing functions as a "motor scheme" or "image of thought."[3] Not only deconstruction but also its predecessors, Heideggerian destruction and Hegelian dialectic, are all "in their own way . . . thoughts of vesperal negation."[4] Each is simultaneously an image of its own decline and the promise of something new emerging. Yet in this regard, as I argue following Malabou's own indication, Malabou's plasticity is no different. She states, "I believe that by following the thread of plasticity it is now possible for me to accede to *another dusk*, or at least to *another meaning of dusk*."[5] It follows that deconstruction must be peculiarly understood both as a predecessor of Malabou's thought and as a form or transformation of the same kind of thought.

The structure of this relationship suggests we may gain insight into both Malabou's relationship with deconstruction and Malabou's thought taken on its own terms by examining deconstruction's means of figuring its relationship with its own predecessors. The following chapter does exactly this, but through an aspect of deconstruction's lineage Malabou does not explicitly discuss: Derrida's relationship with the work of Edmund Husserl. Further, and in line with the spirit of Malabou's notion of the motor scheme, this chapter characterizes these developmental relationships through the transformative potential of two "images of thought" that are not central to Malabou's own writing. First, I develop the notion of "radical responsibility," which Derrida first ascribes to Husserl in his *Introduction* to the latter's "The Origin of Geometry," as a description of plasticity's operation within philosophical writing. In other words, I suggest that responsibility is what plasticity "looks like" as an attribute of the subject who makes transcendental claims about the world.[6] Second, I compare this image of philosophical responsibility to some exemplary Futurist manifestos in order to develop a second image—the political-aesthetic manifesto—this time pertaining to the genre of Malabou's writing. By reading Malabou's *What Should We Do with Our Brain?* as a manifesto, I argue, we can gain a deeper understanding of the radical responsibility underlying her deployment of neuronal plasticity as a motor scheme.

"A RADICAL RESPONSIBILITY": DERRIDA, HUSSERL, AND MALABOU

If Malabou critiques Derrida primarily on the basis of his rejection of form, this rejection is most emphatically rehearsed in Derrida's engagement with

Husserlian phenomenology, which he tends to identify as the epitome of the metaphysical tradition he seeks to overcome. Indeed, both form and vision (the phenomenological ur-sense referred to in the following quotation) are directly implicated in this identification. As Derrida writes in one of *Speech and Phenomena*'s "other essays," "Form and Meaning," "as soon as we use the concept of form—even to criticize another concept of form—we must appeal to the evidence of a certain source of sense. And the medium of this evidence can only be the language of metaphysics."[7]

While these statements appear to indicate a more or less total dismissal of phenomenology—and are taken as such by many commentators—concerns apparent throughout Derrida's thought suggest that his relationship with Husserl is in fact more nuanced, ultimately resembling Malabou's complex relationship with her predecessors. First, several key deconstructive techniques—such as "writing under erasure," in which "inaccurate" metaphysical concepts are nonetheless retained as "necessary" communicative devices in lieu of alternatives[8]—and principles—such as the guideline that deconstruction must unfold wholly within the presuppositions of whatever discourse it dismantles, without employing any external reference points—suggest that the gesture of simple rejection, whether of metaphysics or anything else, is not generally admissible in a rigorous Derridean philosophy. More to the point, it is notable that Derrida's personal attitude toward Husserl remained predominantly positive throughout his career. In Derrida's book-length *Introduction* to Husserl's "The Origin of Geometry," published several years before *Speech*, Derrida ultimately associates Husserl with a kind of "radical responsibility" he deeply admires.[9] Nearly thirty years after the publication of *Speech*, Derrida renews his admiration for this same "responsibility" while referring to Husserl in particular as one of "the great figures in the history of philosophy."[10] Both the logical principles of deconstruction and the tone of Derrida's writing, then, suggest that Husserlian phenomenology and Derridean deconstruction should be viewed according to a commonality or kinship that is as fundamental as their more famous divergence. Without discarding Joshua Kates's argument, in his study of Derrida's early work in *Essential History*, that this kinship is rooted in Derrida's retention of Husserl's distinction between empirical, historical facticity and the region of "irreal" transcendental origins, I propose that the aforementioned "radical responsibility" presents a robust framework with which to track the continuity of phenomenology and deconstruction.

What is Derrida's "radical responsibility"? For Husserl in "The Origin of Geometry," responsibility refers to the phenomenologist's mandate to recover and univocally express the eidetic, transcendental origins that ground phenomena, such as the eponymous "origin of geometry."[11] "Origin" in this case does not refer to a historical, empirical, or sensible origination (e.g., the origin of geometry is not the historical place and time in which geometric science was first practiced), but to the nonsensible, eternal Idea by which a science is exposed to infinite theoretical development. Without disputing the validity of Husserl's mandate to approach such eidetic origins, Derrida understands the full recovery of origins to be essentially impossible, or at least inherently fraught. Interestingly, this is not because the notion of a transcendental origin as such is unviable, since Derrida accepts as valid Husserl's claims to have revealed the origins of many of the "regional sciences" treated in his extended philosophical work. Rather, the problem lies with just one transcendental origin: the Idea of phenomenology itself. According to Husserl's hierarchy, the recovery of this origin is ultimately essential for grounding the entire system of phenomenology and its derivative regional sciences, leading Derrida to refer to this theoretical recovery as phenomenology's *Endstiftung* or "final institution." Yet it is precisely this *Endstiftung* that Husserl never completed and never could have completed; by definition, phenomenology cannot apprehend the Idea of phenomenology. This is because phenomenological intuition, by means of which every origin is apprehended in its being made present to sense, must always presuppose and anticipate its Idea as a regulative possibility at the margins of presence. Although the Idea is operative in every phenomenological intuition, it is never made to appear in the present moment. Thus, the apprehension of its specific content is eternally deferred. As Derrida puts it, the Idea must remain "a hidden sun which shows without being shown."[12]

If the final apprehension of the Idea of phenomenology, its *Endstiftung*, is eternally unachievable, does this mean the phenomenological enterprise is only a chimera, a pipe dream? Actually, the fact that phenomenology takes place despite this unachievability (i.e., that phenomenology, even if it cannot be completed ideally, has a factual history) exemplifies the structure of what Derrida calls "a radical responsibility." Derrida states,

> The *Endstiftung* of phenomenology (phenomenology's ultimate critical legitimation: i.e., what its sense, value, and right tell us about it), then, never directly measures up to a phenomenology. At least this *Endstiftung* can give access to itself in a philosophy, insofar as it is *announced* in a concrete phenomenological evidence, in a concrete *consciousness* which is made *responsible* for it despite the finitude of that consciousness, and insofar as it grounds transcendental historicity and transcendental intersubjectivity. Husserl's phenomenology starts from this *lived anticipation* as a radical responsibility.[13]

Indeed, the Idea of phenomenology is not merely an empty moniker given to an impossible task, although the role of this particular Idea in the phenomenological enterprise differs from that of all others. While phenomenology engages itself theoretically with the recovery of transcendental origins—and almost entirely succeeds in this engagement—the problem of the *Endstiftung* persists as a concrete, factual problem pertaining to phenomenology as a historical consciousness and practice. According to Derrida, "This evidence of the Idea as regulative possibility is absolutely exceptional in phenomenology. . . . It is evidence only insofar as it is *finite*, that is, here, *formal* [formelle], since the content of the infinite Idea is absent and is denied to every intuition."[14]

Derrida's use of the term "formal" here, in one of his earliest texts, is both unique in his writing and extremely suggestive in relation to Malabou's work. Rather than associating form with static presence, as he does in *Speech and Phenomena*, Derrida here links form with finitude, historicity, and movement, aligning his statements with Malabou's understanding of form as implying metamorphosis. Unlike Malabou, however, Derrida's retention of Husserl's transcendental categories leads him to measure the resultant historicity of phenomenological practice against a still operative transcendental ideal. For him, then, this historicity is not a taken-for-granted state of affairs (as it is for Malabou, who claims there can be no outside of history, form, and finitude), but is instead indissociable from the risk of failure that makes responsibility meaningful. That the problem of the *Endstiftung* cannot be solved ensures that phenomenology as such is not transcendentally grounded, and, therefore, that someone must make herself or himself responsible for its concrete outcomes.

Despite this risk of failure, responsibility is not a fault to be abandoned or simply corrected. In a certain sense, Derrida is unequivocal in stating that a philosophy must be coupled to responsibility if it claims to have anything to say about factual reality—in other words, if it is to be a philosophy at all. "Without such a consciousness," he says, "without its own proper dehiscence, nothing would appear."[15] To take responsibility means to take factuality seriously, since the philosopher who takes responsibility for their system premises it on an Idea that can never be grounded in principle but only demonstrated in fact—that is, by living it. But this life (and here we truly mean transcendental life—the life of transcendental ideas, and of the Idea of the transcendental) is by nature exposed to ruination; it is a dying life, a *dusk*. More specifically, a lived philosophy—one for which someone is responsible—must expose itself to what Derrida calls "Absolute Objectivity" or "the origin of Being as History," and the necessity of this exposure makes that philosophy's ruination part of its very existence. At the same time, the persistence of "the origin of Being as History" ensures that there will always be a motivation for philosophers to take on this responsibility, or, in other words, to attempt to answer transcendentally the question Being poses objectively. Again, Derrida describes this structure in a succinct formula:

> to take factuality seriously as such is no longer to return to empiricism or nonphilosophy. On the contrary, it completes philosophy. But because of that,

> it must stand in the precarious openness of a question: the question of the origin of Being as History. Every response to such a question can resurface only in a phenomenological process. Ontology only has a right to the question.[16]

What Derrida ultimately depicts, then, is an asymmetrical economy of manifestation and ruination. If every philosophy, in its own way, makes visible a solution to the question of the origin of Being as History, each of these solutions at once exhibits its own ungroundedness, overextension, and failure, since it is inevitably contaminated by the persistent facticity conferred on it by its target. On the other hand, the specter of this failure also indicates a promise that the question will be asked again, and answered again by a newly responsible philosophical making visible. With this description we come full circle to Malabou's understanding of *dusk*, specifically of philosophies as manifestations of *dusk* and her own philosophical motor scheme as *another dusk*. However, the specific senses of failure and ruination accompanying Derrida's responsibility (thanks to its investment in transcendental ideality) significantly contribute to our understanding of Malabou's plasticity in at least two ways. First, they help clarify the relationship between visibility and invisibility in Malabou's notion of plastic reading as it is described in both *The Future of Hegel* and *Plasticity at the Dusk of Writing*. Second, they help demonstrate the continuity between plastic reading and the neuronal plasticity described in Malabou's neurobiological and psychoanalytic works, especially as it pertains to the privilege Malabou increasingly places on plasticity's capacity for destruction.

Toward the end of *The Future of Hegel*, Malabou argues (through Hegel) that speculative philosophy can only continue via a process of "plastic reading," which closely resembles the economy of philosophical responsibility elaborated in Derrida's work. For Malabou, attempting to apprehend a speculative proposition—for example, "God is being"—involves experiencing "a loss of ground and foundation," since the reader is unable to demarcate a stable transition from the proposition's subject to its predicate.[17] According to Malabou, this results in three distinct movements. First, the reader becomes caught up in the material form of the proposition and its language (which we can compare with Derrida's recognition of the historical facticity of phenomenological practice). Second, "the reader approaches the absolute origin" (which we can compare with Derrida's description of the exposure of the phenomenological process to the origin of Being as History), in whose place she or he discovers *nothing*.[18] Finally, as a result of this encounter with *nothing*, the only means left by which the reader can attempt to "read" the proposition is by rewriting it in a new form (which we can compare with Derrida's description of new phenomenological responses surfacing to answer Ontology's question): Malabou writes that "plunged into the void of the proposition, the reader is brought to formulate new propositions in return."[19]

At the beginning of her "Conclusion," Malabou describes plastic reading both as involving visibility and invisibility at once, and as connoting a "situation of 'in-between.'"[20] That these descriptions, although contained within the same paragraph, are not precisely congruent exemplifies the complex positioning of plasticity and visibility that is also apparent in Malabou's very different description of plastic reading in *Plasticity at the Dusk of Writing*. Sketched explicitly as a response to Derrida's notion of writing, plastic reading in this text refers to revealing or making visible the gap between the traditional metaphysics of presence and the textuality of deconstruction, a gap that deconstruction itself thematizes in terms of the nonvisible and nonformal. In Malabou's words, "It is a matter of causing the form that comes *after presence* to arise in works."[21] What is this "form that comes *after presence*," and to what extent is it comparable either to a simultaneity or an "in-between" of visibility and invisibility? Of course, the insolubility of these formulas is mitigated by Malabou's acknowledgment of their idiomatic quality, as exemplified by her referring to the *anticipatory structure* of Hegelian subjectivity through the French idiom *le 'voir venir'* or "to see (what is) coming."[22] At the same time, if idiom is not just idiom in general but specifically and concretely "one's own idiom," as Malabou presents it through Hegel,[23] this acknowledgment can also direct us to the importance of responsibility.

Derrida's understanding of responsibility, I believe, offers a compelling characterization of Malabou's "form that comes *after presence*." Responsibility, first of all, must adhere to a visible form, which Derrida calls a "concrete phenomenological evidence." Responsibility is concretely *there* in the very existence of the texts of philosophy, such as Husserlian phenomenology; as Kates explains, responsibility characterizes what Derrida (in *Speech and Phenomena*) refers to as the phenomenological voice, which should be taken in both its most transcendental and its most mundane senses.[24] Husserl is responsible for phenomenology not because his name coincides with any invisible or immaterial identity of phenomenological thought (since, again, any such content of the Idea of phenomenology is by definition unknowable), but because he *wrote* phenomenology, and because we can *see* his signature affixed to phenomenological texts. On the other hand, that the very recognition of an act's responsibility is tied to an irrecoverable transcendental Idea means that the act's failure, ruination, or becoming invisible is built into the very fact of its appearing. Finally, then, this copresence of visibility and invisibility in responsibility is clarified by its inherent link to historicity. It is not the case that invisibility as such is somehow made visible by responsibility. Rather, responsibility defines the visible as the history of its own ruination; whatever is "my responsibility" is the visible form of that form's becoming invisible in time. If, then, I am responsible for an image of thought, it is an image of thought *at dusk* and *as dusk*.

In the first place, this envisioning clarifies Malabou's use of metaphors of decline in her explication of Hegel's plastic reading, as she describes its stages as forms of falling, sinking, and plunging. More interesting, though, is the connection between the figure of one's responsibility as an image of its own ruination and Malabou's summative description of the plastic reader's subjectivity as "a substantial accident."[25] We should now be able to see why, over numerous works, Malabou continues to insist that plasticity cannot be understood without incorporating destructive plasticity, which signifies plasticity's ability to destroy form in addition to creating and transforming it. If we were to believe that Malabou's goal is merely to represent the world as an immanent totality of form as metamorphosis, we would be justified in questioning the inclusion of destruction in her system. Why not simply think the world as a positive totality of changing forms, rather than forms destroyed and remade? It turns out, however, that destruction is necessary in order for the very form of this thought—that is, a transcendental and genuinely philosophical thought, one that answers the question posed by Ontology—to appear within the world, and not outside it in a purely transcendent sphere. This is the lesson Derrida extracts from his unflinching belief in the possibility of transcendental thinking: if transcendental thought emerges, someone is responsible. In other words, such thought can only be visible as its own ruination or self-destruction. Further, the fact that thought appears as this kind of image in history—that is, as a motor scheme, the privileged representation of a historical epoch with both a beginning and an end—is what allows for there to be a philosophical tradition at all. Without responsibility, it would be impossible for successive generations to continue to offer solutions to the same transcendental question while renewing these solutions in light of their predecessors' failures. Considering the lineage of dialectic, destruction, deconstruction, and plasticity sketched by Malabou, these conclusions should apply equally well to what she calls speculative thinking in *The Future of Hegel*. The writing of any philosophy, in that it is tied to responsibility, both comes from ruins and is its own ruination, a formalization of its own disappearance.

Considering responsibility's role in explicating the relationship between transcendental subjectivity and its manifestation in the tradition and writing of philosophy, it is unsurprising that the figures of responsibility and destructive plasticity are also central to Malabou's efforts to describe thought and the subject as neurobiological entities. In *Ontology of the Accident*, these efforts are collected in an image closely resembling the formula developed above, wherein thought and the subject each constitute "a form born of the accident, born by accident, a kind of accident."[26] With reference both to the neurobiological structure of the brain and the psychoanalytic structure of the psyche, the essential form of this image is repeated in Malabou's portrait of cerebral suffering in *The New Wounded*. Here, Malabou writes, "The permanent dislocation of one identity forms another identity—an identity that is neither the sublation nor the compensatory replica of the old form, but rather, literally, a form of destruction."[27] My aim, therefore, is not to propose Derridean responsibility as a fuller concept than Malabou's destructive plasticity, nor is it to append responsibility to plasticity as a supplement that would complete or bolster it. Rather, I propose that responsibility and plasticity are two images, two motor schemes, of the same "essential core."[28]

For the same reason, however, the concept of responsibility also offers us a way to live this core differently. If plasticity allows us unprecedented access to the resources of contemporary neuroscience in our investigations of the human subject, responsibility allows us to see that *we* are the ones who think and live as plastic selves, even and especially when we reflect on this situation in transcendental and ontological terms. By balancing the plasticity of responsibility with our and Malabou's responsibility for plasticity—or by treating the two schemes as reversible, in the way Malabou treats structuralism and poststructuralism as reversible[29]—we can better grasp how plasticity adheres beyond the empirically determined world to also condition our transcendental pronouncements about that world.

THE MANIFESTO AND *WHAT SHOULD WE DO WITH OUR BRAIN?*

If Malabou reminds us that no philosophical concept can place itself beyond the exigency of form, our reading of Derrida indicates that the coordinate visibility of the philosophical motor scheme can also be viewed in terms of responsibility, as the formalization of its own ruination. These conclusions point us in the direction of a problem already implicated by Derrida's well-known emphasis on the *writing* of philosophy: the problem of genre. Considering its entanglement with responsibility, what is the genre of Malabou's writing? It should be clear that this writing, to the extent that it formalizes its own ruination in finite time, should not be understood according to the traditional Enlightenment model of a scientific discourse that transparently explicates eternal, objective truths. Seen in terms of its responsibility, no writing achieves a full recovery of its eternal transcendental grounds, but rather holds itself and is held responsible precisely for its failure to achieve this grounding ultimately. It seems reasonable, then, to turn toward genres centered on rhetorical effect and creative expression in classifying Malabou's writing, although we must also attempt to retain an account of how her work manifests in a transcendental philosophical tradition. I propose that the political-aesthetic manifesto, a genre Marjorie Perloff describes as "an art form designed to erase the traditional line between creation and criticism,"[30] fulfills these criteria, and that we can productively read Malabou's work as representative of political-aesthetic manifesto writing. As in the previous section of this chapter, however, I do not intend to present this characterization as the definitive or even best representation of Malabou's work, but to introduce a new motor scheme or image through which her efforts can be interpreted and, more importantly, reproduced and extended.

Malabou's texts have been described as manifestos by commentators such as Clayton Crockett,[31] and Malabou herself refers to *Plasticity at the Dusk of Writing* as a "text-manifesto";[32] nonetheless, these descriptions have not been explored in significant depth. There are, in fact, close resemblances between generally agreed-upon characteristics of the manifesto and the philosophical form of responsibility we have identified with Malabou's writings on plasticity. Most superficially, the manifesto is intimately connected to visibility: the word "manifesto," which varies only slightly across most European languages, is etymologically connected to the English words "manifestation" and "manifest," which means not only "to make public" but also "to make tangible or available to the senses." The manifesto in the Western literary and political traditions is also closely connected to responsibility, especially in the sense of claiming agency and speaking in one's own voice. In all cases, it is centrally important that the content of the manifesto emanates from a certain personal or collective identity, which deploys the text in order to assert its interests and authority against alternatives. At the same time, even though this identity is specific and historically situated, manifestos also generally take the form of universal statements through their frequent and heavy-handed use of the injunctive mood.[33] This absolutism formally connects manifesto writing to the transcendentalism of Derrida's philosophical responsibility.

In "The Rhetoric of the Manifesto," Anne Sinkey combines these themes with a depiction of failure and ruination that clinches the manifesto's connection to Derridean responsibility, as long as we understand Sinkey's political reference to liberalist universality as analogous to Derrida's Husserlian transcendentality. For Sinkey, manifestos are contradictorily rooted in both (1) an individual's or group's desire to be included in universal rights discourse, which is based on a fundamental investment and faith in universality, and (2) a tacit understanding that universality is inherently unachievable, and that there will therefore always be outliers manifesting the desire to be included.[34] The manifesto writer, then, "criticize[s] a vocabulary for being contingent without having anything other than contingency to offer in its place."[35] Yet, since universality is incompletable by definition, the manifesto's specific, historically situated failure is simultaneously a making visible of the universal project's *universal* failure.[36] Thus it is through the making visible of the failure and ruination of a universalist ideal that manifestos give historical form to the absolute. This occurs in the same way that, for Derrida, responsibility *is* the transcendentalism—that is, transcendentalism *in practice*—whose unachievability *in theory* is what makes someone responsible for practicing it.

While many writers note that any given manifesto either emerges from a crisis in the old order or announces and makes manifest such a crisis, the manifesto's connections to philosophical responsibility clarify how this emergence and announcement are in fact the same event: the manifesto, too, is "a form born of the accident, born by accident, a kind of accident."[37] To turn to some privileged examples, this operation is perhaps most visible in the manifestos of the early twentieth-century Futurist movement, which is often seen as inventing the avant-garde genre of the art manifesto.[38] In Filippo Tommaso Marinetti's infamous *The Foundation and Manifesto of Futurism*, the incorporation of ruination and the accident into the very visibility of the manifesto is performed quite literally, as the text narrates how a violent car crash coincides with Marinetti's invention of Futurism's first tenets.[39] Alternatively, the "accident" of the Futurist manifesto can be characterized as its brash refusal to conform logically or teleologically to conventional common sense. This is apparent in the opening statement of Mina Loy's "Aphorisms on Futurism," which effectively condenses the dominant attitudes of both Futurism and the political-aesthetic manifesto in general. In a tight couplet, Loy writes, "DIE in the Past/Live in the Future."[40] It is not hard to see how, considered according to the conventional rules of chronology, this statement fails to describe reality: actually, none of us who are currently alive have died in the past, and we may or may not be alive in the future. Yet it is the fact that these lines have a content—that is, that they work both imagistically and syntactically to constitute the *form* of an actionable pronouncement—despite being generally nonsensical that exemplifies Derridean-Husserlian responsibility or destructive plasticity. Loy's statement is a visible ruination, a failure played out in time: like one of Hegel's speculative propositions, we experience its fall into groundlessness in the very event of reading it. At the same time, to attempt to comprehend this statement means to live it, visibly but also historically, and to be potentially transformed by this act of living that self-destructs.

The manifesto, then, both anticipates a transformed future and enacts the present's transformation into that future, specifically by playing out the failure and ruination of the present moment as such. This operation is integral to what Martin Puchner calls "the futurism effect," by which the manifesto does not only announce the avant-garde artwork to come but actually becomes the defining genre of that artwork.[41] Perloff makes the point clearly in her essay, subtitled "The Manifesto as Art Form," writing, "The novelty of Italian

Futurist manifestos . . . is their brash refusal to remain in the expository or clinical order, their understanding that the group pronouncement, sufficiently aestheticized, can, in the eyes of the mass audience, all but take the place of the promised art work."[42] Later, she drives the point home: "To talk about art becomes equivalent to making it."[43]

If the Futurists' manifesto writing *about* artwork simultaneously *is* that artwork, I propose that Malabou's writing *about* the brain, and specifically the work of the brain, *is* the "brain work" it proposes. This seems to me to be one of the fullest possible interpretations of Malabou's opening statements in *What Should We Do with Our Brain?*: "The brain is a work. . . . We are its subjects—authors and products at once."[44] Despite this, it is easy to mistake *What Should We Do with Our Brain?* for a project of making visible in the Enlightenment sense—that is, as revealing the eternal truth of what has historically been shrouded in ignorance. Superficially, this would seem to be the project suggested by Malabou's repeated assertion that "we do not know" the plasticity of the brain she herself posits, for example, when she writes, "Our brain is plastic, and we do not know it. We are completely ignorant of this dynamic, this organization, and this structure."[45]

However, if we commit to viewing these statements as responsible statements—that is, as implicated in the historical facticity they describe and not transcendently isolated from it[46]—we can recognize an effect of Malabou's writing closely resembling that of Loy's political-aesthetic manifesto. How is it that Malabou can tell us "our brain is plastic," then immediately note that "we" are ignorant of what she knows and has just told us? The arc of this statement, like the many similar statements that populate the opening of *What Should We Do with Our Brain?*, follows the trajectory of Derrida's philosophical responsibility. Like Husserl's recovery of transcendental origins, Malabou's concept of plasticity aims to certify a grounded link between, on one hand, the various concrete factual events of history, and, on the other, an eternal Idea or origin underlying this variety: the brain. Malabou asserts that "the structural bond here is so deep that in a certain sense it defines an *identity*. It's not just that the brain has a history . . . but that it *is* a history."[47] Yet if Derrida's analysis of the inherent unachievability of phenomenological grounding is taken seriously, Malabou's intention to "awaken a consciousness of this history"[48] cannot be seen as a final awakening to eternal truth. Rather, Malabou's insistent reiteration of the slogan "we do not know it" indicates that this ignorance or incompleteness of our knowledge is inseparable from her transcendental identification of the brain as the origin of history. Despite the extent to which we can come to know that the brain *is* its history—that is, despite the extent to which such a transcendental notion can be formalized or made visible—this knowledge is also by definition incomplete, and we cannot in fact know it as certainly as the form of the proposition suggests. Malabou's proposition is responsible for the lived experience of this disjunction, which manifests as a making visible of its own descent into transcendental ruin.

The inherent failure of Malabou's gesture is, in fact, perfectly consistent with Malabou's depiction of plasticity. This is because plasticity, since it incorporates destructive plasticity in addition to its other forms, is "not only the creator and receiver of form but also an agency of disobedience to every constituted form, a refusal to submit to a model."[49] The notion that "the brain is its history," or that it is plasticity, cannot help but represent another model to which plasticity would refuse to submit. But it is precisely for this reason that the proposition "the brain is its history" is also a concrete formalization of the brain's becoming historical: the proposition is an image or motor scheme through which thought comes to embody a particular epoch—in this case, that of neuronal man—which, since it is historical, incorporates its own coming to an end. As we saw in Derrida's depiction of the philosophical tradition as a tradition of responsibility, it is only because each motor scheme incorporates its own ruination that it can emerge from its predecessors' failures. In other words, to emerge from the *dusk of writing*, plasticity must be *another dusk* already metamorphosing into some other form. At the same time, as we saw in the "futurism effect" of the political-aesthetic manifesto, this instability and self-destruction also gives plasticity the power to *be* the revolutionary "brain work" it calls for. In Malabou's words, "it is precisely because—contrary to what we normally think—the brain is not already made that we must ask what we should do with it."[50] Only the incompleteness of the motor scheme, its making visible its own ruination, can both activate its transformative potential—in other words, its plasticity—and affirm our agential role in putting this transformative potential to work, making us responsible for a brain that is truly ours.

NOTES

1. Catherine Malabou, *Plasticity at the Dusk of Writing: Dialectic, Destruction, Deconstruction*, trans. Carolyn Shread (New York: Columbia University Press, 2010), 49; Catherine Malabou, *La Plasticité au Soir de l'Écriture: Dialectique, Destruction, Déconstruction* (Paris: Éditions Léo Scheer, 2005), 93.

2. Malabou, *Plasticity at Dusk*, 52.

3. Ibid., 14. A "motor scheme," discussed in pages 12–15 of *Plasticity at the Dusk of Writing*, is an image or group of images through which a thought or concept manifests as the privileged representation of a historical epoch. Malabou notes that, over the course of this manifestation, the "narrow lens" of the motor scheme also necessitates a certain mutation or metamorphosis of the thought's "essential core"

(ibid., 13). In her words, "To think is always to schematize, to go from the concept to existence by bringing a transformed concept into existence" (ibid.).

4. Ibid., 17.

5. Ibid.

6. Although Malabou discusses Derrida's conception of responsibility in "An Eye at the Edge of Discourse" (trans. Carolyn Shread, *Communication Theory* 17 [2007]: 22–23), she does not explicitly elaborate the connection between responsibility and plasticity. Further, while Malabou's discussion foregrounds the responsible subject's visibility via the structure of the gaze, her definition does not extend to the formal manifestation of that for which the subject is responsible. In general, I hope that my depiction of responsibility through the remainder of this chapter develops these connections while still tacitly conforming to the characteristics Malabou outlines.

7. Jacques Derrida, *Speech and Phenomena, and Other Essays on Husserl's Theory of Signs*, trans. David B. Allison (Evanston, IL: Northwestern University Press, 1973), 108.

8. Gayatri Chakravorty Spivak, translator's preface to *Of Grammatology*, by Jacques Derrida, corrected ed. (Baltimore: Johns Hopkins University Press, 1997), xiv.

9. Jacques Derrida, *Edmund Husserl's Origin of Geometry: An Introduction*, trans. John P. Leavey Jr. (Lincoln: University of Nebraska Press, 1989), 141.

10. Jacques Derrida, "Remarks on Deconstruction and Pragmatism," in *Deconstruction and Pragmatism*, ed. Chantal Mouffe (London: Routledge, 1996), 83–84.

11. Edmund Husserl, "The Origin of Geometry," trans. David Carr, in *Edmund Husserl's Origin of Geometry: An Introduction*, by Jacques Derrida (Lincoln: University of Nebraska Press, 1989), 165.

12. Derrida, *Introduction*, 144.

13. Ibid., 141.

14. Ibid., 139; Jacques Derrida, introduction to *L'Origine de la Géométrie*, by Edmund Husserl (Paris: Presses Universitaires de France, 1962), 153.

15. Derrida, *Introduction*, 153.

16. Derrida, *Introduction*, 151.

17. Catherine Malabou, *The Future of Hegel: Plasticity, Temporality and Dialectic*, trans. Lisabeth During (London: Routledge, 2005), 177.

18. Ibid., 179.

19. Ibid.

20. Ibid., 184.

21. Malabou, *Plasticity at Dusk*, 57.

22. Malabou, *The Future of Hegel* 13. Malabou explains that the expression *voir venir* can "refer at one and the same time to the state of 'being sure of what is coming' ('être sûr de ce qui vient') and of 'not knowing what is coming' ('ne pas savoir ce qui va venir'). It is on this account that the 'voir venir,' 'to see (what is) coming,' can represent that interplay, within Hegelian philosophy, of teleological necessity and surprise"—or, in an analogous context, of visibility and invisibility (ibid.).

23. Ibid.

24. See the fifth chapter of Kates's *Essential History*, "Husserl's Circuit of Expression and the Phenomenological Voice in *Speech and Phenomena*." In the book's introduction, Kates also writes that "[t]he phenomenological voice . . . is what structurally allows sense to stay close to itself, even as it passes beyond itself (toward the infinity of ideality). . . . At the same time, this voice, as Derrida ultimately analyzes it, remains a *phenomenon*" (*Essential*, xxvi).

25. Malabou, *The Future of Hegel*, 183.

26. Catherine Malabou, *Ontology of the Accident: An Essay on Destructive Plasticity*, trans. Carolyn Shread (Cambridge: Polity, 2012), 2.

27. Catherine Malabou, *The New Wounded: From Neurosis to Brain Damage*, trans. Steven Miller (New York: Fordham University Press, 2012), 18.

28. Malabou, *Plasticity at Dusk*, 13.

29. Catherine Malabou, "Following Generation," trans. Simon Porzak, *Qui Parle* 20, no. 2 (2012): 31.

30. Marjorie Perloff, "'Violence and Precision': The Manifesto as Art Form," *Chicago Review* 34, no. 2 (1984): 66.

31. Clayton Crockett, foreword to *Plasticity at the Dusk of Writing: Dialectic, Destruction, Deconstruction*, by Catherine Malabou (New York: Columbia University Press, 2010), xi.

32. Malabou, "Following Generation," 30.

33. Claude Abastado, "Introduction: À l'Analyse des Manifestes," *Littérature* 39 (1980): 9.

34. Anne Sinkey, "The Rhetoric of the Manifesto" (PhD diss., Emory University, 2009), 8.

35. Ibid., 20.

36. Ibid., 4.

37. Malabou, *Ontology of the Accident*, 2.

38. More substantially, Futurism first combined revolutionary political ideals, evident in their explicitly taking up the term "manifesto," with the creative techniques of the *fin de siècle* literary-aesthetic manifesto writers (whose texts were not explicitly called manifestos by their authors). Benedikt Hjartarson, "Myths of Rupture: The Manifesto and the Concept of Avant-Garde," in *Modernism*, ed. Astradur Eysteinsson and Vivian Liska (Amsterdam: John Benjamins, 2007), 177.

39. Filippo Tommaso Marinetti, *The Foundation and Manifesto of Futurism*, in *100 Artists' Manifestos: From the Futurists to the Stuckists*, ed. Alex Danchev (London: Penguin, 2011), 4.

40. Mina Loy, "Aphorisms on Futurism," in *100 Artists' Manifestos: From the Futurists to the Stuckists*, ed. Alex Danchev (London: Penguin, 2011), 63.

41. Martin Puchner, *Poetry of the Revolution: Marx, Manifestos, and the Avant-Gardes* (Princeton, NJ: Princeton University Press, 2006), 93.

42. Perloff, "Violence," 70.

43. Ibid., 74.

44. Catherine Malabou, *What Should We Do With Our Brain?*, trans. Sebastian Rand (New York: Fordham University Press, 2008), 1.

45. Ibid., 4.

46. Malabou notes that "we are not seeking to explain or explicate consciousness, but to *implicate* it" (ibid., 11).
47. Ibid., 1.
48. Ibid., 2.
49. Ibid., 6.
50. Ibid., 7.

BIBLIOGRAPHY

Derrida, Jacques. *Edmund Husserl's Origin of Geometry: An Introduction*. Translated by John P. Leavey, Jr. Lincoln: University of Nebraska Press, 1989.

———. Introduction to *L'Origine de la Géométrie*, by Edmund Husserl, 3–171. Paris: Presses Universitaires de France, 1962.

———. "Remarks on Deconstruction and Pragmatism." In *Deconstruction and Pragmatism*, edited by Chantal Mouffe, 79–90. London: Routledge, 1996.

———. *Speech and Phenomena, and Other Essays on Husserl's Theory of Signs*. Translated by David B. Allison. Evanston, IL: Northwestern University Press, 1973.

Hjartarson, Benedikt. "Myths of Rupture: The Manifesto and the Concept of Avant-Garde." In *Modernism*, edited by Astradur Eysteinsson and Vivian Liska, 173–94. Amsterdam: John Benjamins, 2007.

Husserl, Edmund. "The Origin of Geometry." Translated by David Carr. In *Edmund Husserl's Origin of Geometry: An Introduction*, by Jacques Derrida, 157–80. Lincoln: University of Nebraska Press, 1989.

Kates, Joshua. *Essential History: Jacques Derrida and the Development of Deconstruction*. Evanston, IL: Northwestern University Press, 2005.

Loy, Mina. "Aphorisms on Futurism." In *100 Artists' Manifestos: From the Futurists to the Stuckists*, edited by Alex Danchev, 62–66. London: Penguin, 2011.

Malabou, Catherine. "An Eye at the Edge of Discourse." Translated by Carolyn Shread. *Communication Theory* 17 (2007): 16–25.

———. "Following Generation." Translated by Simon Porzak. *Qui Parle* 20, no. 2 (2012): 19–33.

———. *The Future of Hegel: Plasticity, Temporality and Dialectic*. Translated by Lisabeth During. London: Routledge, 2005.

———. *The New Wounded: From Neurosis to Brain Damage*. Translated by Steven Miller. New York: Fordham University Press, 2012.

———. *Ontology of the Accident: An Essay on Destructive Plasticity*. Translated by Carolyn Shread. Cambridge: Polity, 2012.

———. *La Plasticité au Soir de l'Écriture: Dialectique, Destruction, Déconstruction*. Paris: Éditions Léo Scheer, 2005.

———. *Plasticity at the Dusk of Writing: Dialectic, Destruction, Deconstruction*. Translated by Carolyn Shread. New York: Columbia University Press, 2010.

———. *What Should We Do With Our Brain?* Translated by Sebastian Rand. New York: Fordham University Press, 2008.

Marinetti, Filippo Tommaso. *The Foundation and Manifesto of Futurism*. In *100 Artists' Manifestos: From the Futurists to the Stuckists*, edited by Alex Danchev, 1–8. London: Penguin, 2011.
Perloff, Marjorie. "'Violence and Precision': The Manifesto as Art Form." *Chicago Review* 34, no. 2 (1984): 65–101.
Puchner, Martin. *Poetry of the Revolution: Marx, Manifestos, and the Avant-Gardes*. Princeton, NJ: Princeton University Press, 2006.
Sinkey, Anne. "The Rhetoric of the Manifesto." PhD diss., Emory University, 2009.
Spivak, Gayatri Chakravorty. Translator's Preface to *Of Grammatology*, by Jacques Derrida, corrected edition, ix–lxxxvii. Baltimore: Johns Hopkins University Press, 1997.

Section 2

ARE NEW ATTACHMENTS POSSIBLE?

On Habit and Habitual Returns

Chapter 4

Habitual Propensity: Plastic or Elastic?

An Encounter between Catherine Malabou and Sigmund Freud on the Phenomenon of Habit[1]

Sandrine Hansen

In the conclusion of *The Future of Hegel* Catherine Malabou's philosophical trajectory crystalizes around the phenomenon of habit. At this moment where impasse and projection converge, habit arises as the epitome of both the challenge and the future of philosophy. Malabou writes,

> From now on we can no longer have anything to do with things other than our own habits, in which we find ourselves immediately alienated. Thought's very life depends on its power to awaken that vital energy which always tends to "mortify" itself, to become sedimented into fixed and rigid positions. The outcome that will follow depends on its awakening: thought has nothing to do but wait for the habitués to look at their habits.[2]

If the future of thought, and philosophy, in so far as its labor consists in thinking, depends on our capacity to question habit, we are faced with a challenge. Malabou and Slavoj Žižek, following in Jacques Derrida's wake, have classified habit as a "pharmakon," both a remedy and poison.[3] Furthering our existence with spontaneous reactions habit facilitates our everyday life, but it may also, as many of us surely knows, throw us back into inappropriate and even dangerous actions before we have time to reflect. To think about habit requires therefore that we grasp or conceptualize this *pharmakological* nature. Habit is, as Hegel wrote, "open to anything we chance to put into it,"[4] vital or not. This notorious promiscuity of the phenomenon of habit is one aspect of its *pharmakological* nature, another, which will be the focus of this chapter, is the propensity to repeat, the force of habit.

Thomas Reid drew a distinction between the facility and the propensity or disposition of habit. Considering habit as a principle of action, Reid held that habit should be defined on the basis of the causal force with which it impels us to act, that is, the propensity of habit to repeat. Habit therefore should not be characterized and defined in consideration of the effortlessness or facility with which we undertake frequently repeated actions, but in regard to the effort it takes to keep them in check.[5] Undoubtedly there is something uncanny about the propensity of habit. The seemingly mechanistic push with which our actions, and even our thoughts, repeat themselves betrays the presence of another, an automaton lurking in our most idiosyncratic acts. The contrast between the freedom of the subject and this untouchable other served, it appears, in the predominant modern interpretation as a premise for bracketing the ontological significance of habit.[6] Against this background the objective of this chapter is to investigate how we can conceptualize this seemingly mechanistic propensity of habit in consideration of the phenomenon's *pharmakological* nature and its ontologically constitutive importance.

To investigate this persistent propensity to repeat, supposedly at work in all creatures of habit, I will address the phenomenon of compulsive repetition. Although compulsive repetition is a phenomenon somewhat besides the commonsense understanding of habit insofar as it might entail too strong an emphasis on the phenomenon's pathological peril, this approach is justified by the scope of the present investigation. If we cannot acknowledge today that the human being remains burdened by the persistence of futile habits that threaten our very livelihood, individually as well as collectively, we may start to wonder if we will ever be able to ask the right questions concerning habit, addiction, obsession, and other conservative human phenomena. In addition, we should not forget that the conceptual distinction between habit and addiction is itself a relatively new construction.[7] The ramifications of this area of research are vast; here I will focus solely on how possibly to conceptualize the propensity of habit to repeat in consideration of the constitution of the self.

As one attempt to begin to think through the *pharmakological* nature of habit, this chapter challenges and nuances the dominating philosophical interpretations of habit as a phenomenon of plasticity.[8] William James, who explicitly drew a connection between our organic plasticity and the phenomenon of habit, defined plasticity as the "possession of a structure weak enough to yield to an influence, but strong enough not to yield all at once."[9] On this basis, it appears evident that organic plasticity is a precondition for contracting habits. Nevertheless, it is questionable to what extent the conceptualization of habit as an instantiation of plasticity, or "open-ended plasticity" as Elizabeth Grosz recently put it,[10] can account for the nonadaptive compulsive force of habit. Traditionally, the concept of plasticity is defined as the capacity to receive and give form. Malabou, however, reminds us of the negative work of plasticity. "Plastic," she writes, "is the name of an explosive material. Plasticity may be used to describe the crystallization of form as well as the destruction of all form (as suggested by the term *plastic explosive* for a bomb)."[11] In continuation of her work on the annihilating capacity of plasticity Malabou introduces a theory of the plasticity of compulsive repetition. Focusing on the compulsive character of habit I will juxtapose Malabou's concept of the plasticity of compulsive repetition with Freud's understanding of compulsive repetition as a manifestation of a kind of organic elasticity. Whereas Malabou's destructive plasticity harbors a creation of something new, the organic elasticity Freud sees manifested in compulsive repetition is rather a rebound effect, a force throwing the individual back upon an earlier organization. The juxtaposition of Malabou's and Freud's conception of compulsive repetition uncovers a metaphysical question at the heart of habit: is the adhesiveness and the tendency to repeat manifested in the phenomenon of habit an inexhaustible, ontological potential, or is it, on the contrary, the expression of an inherent inertia resisting change?

But before we can arrive at these metaphysical considerations I will begin by outlining the distinction between facility and propensity in consideration of a more general characterization of habit.

HABIT AND THE PRINCIPLE OF REVERSIBLE ENERGIES

There are many ways to approach the phenomenon of habit. One is, as Malabou points out, to look at the "principle of reversible energies" governing habit. Malabou explains, "Habit is at first an effect, a way of being that *results* from change, but gradually becomes a *cause* itself, as it initiates and maintains repetition."[12] In the process of habit formation, what was contingent—a specific flavor, a bus schedule, an accidental encounter—becomes essential to the habituated being.[13] Habit thus manifests a form of inversion in which the external world is folded inward. Repeated excitations do not provoke agitation; they are rather constitutive of the "familiar." Our everyday conduct seamlessly unfolds in accordance with the social demands and the environment to which we are accustomed. The strength of the transposition in which contingent affects are elevated to necessary structures is so strong that we may even wonder whether the experience of transgression and uncanniness have their roots in the phenomenon of habit.

As the consequence of change, the phenomenon of habit conserves what has been imparted to the being. According to James, habits are drawn like pathways in our nervous system, by practice or repetition. As we repeatedly tread these pathways they become fixed, they "set like plaster."[14] Habit is thus, as suggested in the passage by Malabou above, perceived as a fossilization or sedimentation of spirit or life.[15]

The metaphor of pathways explains the facility of habit, but habit is also a cause of change; it has a particular momentum. Following Hume, this change is the foundation of our experience of cause and effect. Without habit we would be incapable of inferring from experiences, and would remain inept captives of the present, lacking the ability to create a life for ourselves and unable to avoid approaching dangers. Thus, according to Hume, habit is "the great guide of human life. It is that principle alone, which renders our experience useful to us, and makes us expect, for the future, a similar train of events with those which have appeared in the past."[16] The repetition of an effect or an operation generates "a propensity to renew the same."[17] It is the contraction of this propensity that allows us to project ourselves into the future on the basis of customary conjunction. Hume describes the result of habit formation—itself also a cause—as a nonreflective difference in sentiment, something we know best perhaps, from what we refer to as the weight of expectation. Deleuze finds in this difference of sentiment the key to the paradox of repetition: "habit *draws* something new from repetition—namely, difference."[18]

For Hume, the ultimate reason behind the propensity to repeat, effected by habit, remained an epistemological threshold for philosophy.[19] Félix Ravaisson to the contrary took the enigma of the propensity to repeat as an invitation to forge his spiritualist metaphysics.[20] In Ravaisson's view, it is not action or the concentration of intellectual effort that generates the unconscious inclination to repeat, nor is it a physical change; it is rather an obscure and unreflective tendency generating beneath the experience of the self.[21] Habit, addictions, and other forms of compulsive repetition are, according to Ravaisson, the result of a gradual intensification of an obscure inner force, which rises like an electrical charge or a tone, to counteract and neutralize the feeling of effort and external agitation. This implies that craving a cigarette, for instance, is not the expression of a lack on the side of the subject. The anticipation contained in the craving is rather a manifestation of a surplus of internal activity meant to counteract the agitation of nicotine.[22] In opposition to the conception of habit as a fossilization of life, Ravaisson understands habit as an exaltation of the spontaneity of desire or spirit which permeates all being, in a perpetual contraction of ideality and existence.[23] In Ravaisson's spiritualist ontology, it is not an alien automaton that drives our seemingly mechanical reiterations, but the "unrepresentable dynamism of life."[24] Though Ravaisson clings to a form of Aristotelian teleology, it is difficult in consideration of the examples he chooses—tics, addiction, convulsions, and so on—to see how this inexhaustible force of life is univocally in the service of the individual organism. After all, in Ravaisson's theory of habit, it is the force of life that roars in our every craving, cravings that continuously impel us for better or worse to exceed and expand the existing limits of our being.

Ravaisson's conceptualization of habit stresses the generative aspect of repetition, repetition accumulates and intensifies the dynamism of life. There is no such thing as sterile repetition in his conception of habit. But with the classification of habit as a pharmakon, we must also come to terms with the sterile compulsion of habit that persevere beyond the redundant.

THE PLASTICITY OF COMPULSIVE REPETITION

As soon as we begin to question habit within the theoretical framework of psychoanalysis, we see just how difficult it can be to account for the phenomenon of repetition, and, in this case, the compulsive perseverance of habit. Throughout his metapsychological writings, though especially in *Beyond the Pleasure Principle*, Freud struggled to explain the phenomenon of compulsive repetition. Freud confronted strange cases of repetitive behaviors that yielded no advantage, relief, or pleasure, hence the question: Is there a beyond the pleasure principle, or do all mental phenomena, including unpleasurable repetitions, work in the service of the pleasure principle with the objective of reducing or stabilizing excitation? Malabou's analysis is definitive. Freud never manages to consolidate the existence of a beyond the pleasure principle, and consequently he fails to give a form to the phenomenon of compulsive repetition.[25] The compulsion to repeat, she argues, remains a propaedeutic, a preface, to pleasure. Binding the excessive free energy of a traumatic blow (whether physical or not), the compulsion to repeat enables the proper functioning of the pleasure principle, and is thus subsumed under the dominance of the pleasure principle.[26]

Malabou captures the problem and enigma of repetition, particularly as it occurs in Freud's metapsychological writings, in the following way:

> The profound ambiguity of repetition thus inheres in its binding power. Certainly, in an essential respect, this power is mortiferous: it immobilizes, freezes, or leads to inertia and to the inorganic state. Compulsion—as has been said and resaid—has the spectral character of a death machine. At the same time—something said less often—this "mechanicity" is a *binding agent*: it disciplines, flattens, and tames as it immobilizes.[27]

In *Beyond the Pleasure Principle*, Freud considers different cases which appear to contradict the dominance of the pleasure principle. One of these cases is the compulsive repetition of traumatized individuals. Freud defines a traumatic event as an instance where an external excitation is so strong that it breaks through the protective shield that must prevent an increase of tension in the organism.[28] Flooded by external stimuli, the organism must first master the overwhelming influx, before the pleasure principle, which has been put temporarily out of action, can regain precedence.[29] While dreaming had been understood functionally by means of wish fulfillment as a guard against unpleasurable tensions disturbing one's sleep, the nightmares accompanying war trauma posed a curious challenge. Freud discovered that individuals suffering from a traumatic neurosis repeatedly find themselves thrown back upon the traumatic event in dreams. To make sense of this Freud drew on Joseph Breuer's distinction between a bound cathexis and a freely moving cathexis, suggesting that this unpleasurable unconscious repetition enables the subject to retrospectively master or tame the excessive stimuli, thus making it possible for the subject to "dispose" or rid himself of the overwhelming energy of the incident.[30] Consequently, this compulsion to repeat the traumatic event does not contradict the pleasure principle but appears, Freud writes, to work "independent[ly] of it and seems to be more primitive than the purpose of gaining pleasure or avoiding unpleasure."[31]

At the crux of Malabou's critique of Freud is the phenomenon of binding. The problem of associating the process of binding or mastering with the compulsion to repeat[32] is that one thereby "exhausts the plasticity of free energy, restricts this energy's power of formation and transformation, and thereby obliterates the hypothesis of the plasticity of death."[33] Reducing the compulsion to repeat to a "binding agent"[34] implies that the touch of death is bereft of its formative and transformative power, out of which, according to Malabou, a new (posttraumatic) individual may emerge.[35] It is nevertheless evident, Malabou writes, that there is a beyond the pleasure principle. In traumatized individuals, we see this as the work of destructive plasticity. The work of destructive plasticity is the "formation of death in life, the production of individual figures that exist only within the detachment of existence."[36] These individuals, cool and detached from the life world of their former selves, repeat the traumatic event in an indifferent manner. Consequently, we must "acknowledge that there is *a plasticity of the compulsion to repeat.*"[37] Malabou explains, "What I do wish to underscore is that reiteration entails a power of transformation, and that this power is defined less by the production of differences than by *the possibility of not binding repetition.*"[38] By introducing a scission between repetition and binding, it is Malabou's intention to bracket the synthetic potential of repetition, inhibiting a formal conceptualization of *a* beyond the pleasure principle. What is characteristic about the compulsion to repeat is that there is a cleavage between repetition and that which is repeated.[39] Spinning in its own vortex, the "plastic compulsion" to repeat is its *own* work.

It is clear that Malabou has a particular end in view when she presents us with her idea of the plasticity of compulsive repetition. In opposition to the dominant, positive conceptualization of plasticity Malabou's work on destructive plasticity is a response to the bleak echo of the Spinozistic insight that we do not know what our bodies can do.[40] Hence, "destructive plasticity reveals the possibility, inscribed within each human being, of *becoming someone else at any moment*."[41] Malabou ascribes the indifferent repetition, the repetition without pleasure, to those suffering from cerebral trauma and to soldiers returning from war with posttraumatic stress disorder (PTSD). In the case of these individuals, an event has disrupted the meaningful continuity of their lives. Constituted in the very reception of a shock, the compulsive repetition is in Malabou's words "the mimetic reappropriation of traumatic passivity."[42] On this view, the compulsion to repeat manifests the limit of adaptive reappropriation; it remains meaningless, but it is, Malabou insists, creative of a form.

THE INDIFFERENCE OF HABIT

Returning now to the phenomenon of habit, we must however ask whether the possibility of a nonbinding or nonsynthesizing repetition, defined as the plasticity of the compulsion to repeat, can further our understanding of the *pharmakological* nature of habit. With an implicit reference to Deleuze, Malabou writes that what we need today is a book called *Indifference and Repetition*.[43] While it might be too strong a claim to say that by bracketing the synthetic potential of repetition, the plastic work of compulsive repetition emerge at the expense of difference as such, which as we know according to Deleuze would make repetition itself impossible, it is nevertheless Malabou's intention to highlight that the result of this repetition is not a constructive difference that constitutes a productive relation between the organism and the environment, but an effect rather characterized as *in*difference. Indifference, understood as a psychological attitude, is for Malabou a cold-blooded leveling; it is the annulment of emotional differences and affective relief of the life world.[44] The indifferent subject is no longer affected, not even by her meaningless repetition.

If we turn to Deleuze, the idea of a repetition without binding cannot be aligned with the phenomenon of habit. Contracting a habit is the binding of a difference, resulting from passive synthesis of experiences.[45] Hume pointed out that it is through habit formation that we acquire the expectation that the future will resemble the past. This expectation grounding our experience of causal relations is the so-called difference produced by repetition. Habit as such is what enables us to project and sustain a stable reality.[46] Yet habit is also a phenomenon that conceals its own origin and leaves us unaffected or indifferent to our familiar habitat, thus, even if we accept that habit is a contraction of a difference (in sentiment), it is clear that habit is also a phenomenon in which we are left indifferent to this difference, indifferent in our very familiarity with the reoccurrence of the same. It is in this intertwinement of an affective difference contracted and interiorized as an effect in the being and the indifference this change generates in relation to itself that the ambiguous character of habit comes to the fore. Here Malabou's understanding of the compulsion to repeat appears to offer us some insight into the *pharmakological* nature of habit.

Habit effectively levels out the affective disturbances our environment could initially provoke and if granted enactment, habit even conceals its own propensity. From the perspective of Malabou's theory of compulsive repetition, the indifference of habit suggests that in this mundane phenomenon a cleavage opens between what is repeated and repetition itself. Seeing that this compulsion to repeat continues disregarding what it is that is repeated openly challenges the understanding of habit as an adaptive phenomenon. If the compulsion to repeat, as Malabou writes, is as an immanent possibility inscribed in each of us, the split between what is repeated and the repetition itself could in the case of habit perhaps be grasped as the manifestation of an inherent tension between the subject's capacity to receive and to give form. But we must not forget that Malabou's positive conceptualization of compulsive repetition is bound to an event, and that it is associated with the "existential improvization" that arises in a traumatic accident or a neurodegenerative disease.[47] Compulsive repetition, then, for Malabou is the manifestation of a power of transformation, creating new figures or subjects that "exists only within the detachment of existence."[48] While it is worthwhile to discuss in what way destructive plasticity can positively be considered a "power," the question we confront in consideration of the phenomenon of habit, is whether the compulsion of habit can really be understood as a creative ontological potential manifesting a capacity for change, adaptive or nonadaptive, or whether this persevering compulsion is better understood in terms of an original resistance to change. To explore the latter position, I will now turn to Freud's idea of organic elasticity.

ORGANIC ELASTICITY AND THE COMPULSION OF DESTINY

On the basis of his therapeutic observations, Freud writes in *Beyond the Pleasure Principle* that the "compulsion to repeat . . . exhibit[s] to a high degree an instinctual [*triebhaft*] character."[49] This is a perplexing observation because, first, it seems to suggest that there is something in the instincts that eludes the pleasure principle, and, second, because it contradicts the usual understanding of an instinct as something "impelling [the organism] toward change and development."[50] Freud therefore asks, "How is the predicate 'instinctual' related to the compulsion to repeat"?[51] In reply he writes, "*It seems, then, that an instinct is an urge inherent in organic life to restore an earlier state of things*, which the living entity has been obliged to abandon under the pressure of external disturbing forces; that is, it is a kind of organic elasticity, or, to put it another way, the expression of the inertia inherent in organic life."[52] Unlike plastic materials that retain their deformations and cannot return to their earlier form, the concept "elasticity" designates a tendency or capacity to return to an initial form after undergoing deformation.[53] Freud uses the concept "plasticity" to designate the healthy fluidity of the libido, and the indestructibility of psychic life, but there is, Malabou writes, no *plastic work of the death drive*.[54] In Malabou's view, Freud's characterization of the inherent inertia of the drives as elastic, rather than plastic, introduces an antithesis exactly at the temporal simultaneity of death and life, making it impossible to grasp the undecidable coexistence of life and death in every organism.[55] This is why, Malabou claims, Freud fails to find *a* beyond the pleasure principle that would enable him to give the compulsion to repeat a form and a face.[56]

Whereas Malabou stresses the absolute ingenuity of the destructive plasticity ("a wound gives rise to a *new person*"),[57] she critiques Freud for clinging to the idea of something original within the individual that resists change. But in his therapeutic work Freud comes across something interesting—he is faced with reoccurrences of what he refers to in *Analysis Terminable and Interminable* as fragments of "necrotic bone."[58] Comparing the persistent pathogenic material to necrotic bone seems to suggest that Freud, in cases like unresolved transference and compulsive repetition, caught sight of something that cannot be adequately reduced to a conflict in the individual's psychological development. A disturbance that repeatedly resurfaces like a fragment of necrotic bone in compulsive circularity, disrupting the linear unfolding of organic life, is nothing less than an inorganic deadweight in the midst of life. The phenomenon of compulsive repetition, characterized by Freud as an instance of organic elasticity, is the vacuous echo of an untouchable residue of something dead already within life.[59] Freud writes,

> The elementary living entity would from its very beginning have had no wish to change; if conditions remained the same, it would do no more than constantly repeat the same course of life. . . . Every modification which is thus imposed upon the course of the organism's life is accepted by the conservative organic instincts and stored up for further repetition. Those instincts are therefore bound to give a deceptive appearance of being forces tending towards change and progress, whilst in fact they are merely seeking to reach an ancient goal by paths alike old and new.[60]

It is interesting to note how easily the dominating conception of habit as an adaptive phenomenon can be aligned with Freud's characterization of the conservative instincts of the organism. Whereas the capacity to receive and give form is traditionally associated with the plastic and adaptive character of habit, the tendency to repeatedly return to these earlier externally enforced modifications is seen here as the effect of a conservative obstinacy within the organism.

The compatibility of the two views rests on the notion of acceptance.[61] In the passage above, Freud does not problematize the individual's capacity to accept modifications. However, in his *Analysis Terminable and Interminable* this is thematized in his examination of cases where psychoanalytic therapy appears to make no headway. While he acknowledges that analysis is expected to come up against a certain amount of psychical inertia, there are cases that can only be explained as the manifestations of a "depletion of plasticity."[62] In such cases, the subject's mental processes remain fixed and rigid, making it impossible for them to succumb to the suggestion of analysis. Here it is curious to note that Freud associates the depletion of plasticity with the force of habit. Explaining the cases in which subjects show an insurmountable resistance against the progress of analysis, Freud writes: "One finds the same thing in very old people, in which case it is explained as being due to what is described as force of habit or an exhaustion of receptivity."[63] The correlation between the effect of an exhaustion of receptivity and the force of habit suggests that although the adaptive, or constructive face of habit, that is, the reception of form and facilitation can, as we saw above, be aligned with the tendency of the conservative instincts, the organic elasticity underlying the compulsive repetition would have to be considered both the cause and explanation of the propensity of habit as a rebound effect, while in the same time be understood as an inherent limit for the contraction of new habits (i.e., the depletion of plasticity).

The radical position Freud takes is evident when he writes that the organism struggles to maintain its life, *only to die in its own way*. Even the instincts of self-preservation and self-assertion have as their objective to "ward off any possible ways of returning to inorganic existence other than those which are immanent in the organism itself."[64] Not only are apparent phenomena of change and adaption related to an inherent resistance to change, the struggle for life is itself merely a side effect of organic elasticity, or what Freud surprisingly named the death drive.[65] Although the obscurities of the relation between the death drive and the pleasure principle cannot be resolved easily, this indicates that although we can in certain cases interpret the work of compulsive repetition as a preface to pleasure due to its binding capacity, as Malabou stresses in her analysis, Freud sees in the phenomenon of compulsive repetition something that falls outside, or *beyond* the dominance of the pleasure principle, understanding its binding character only as a secondary effect.[66]

It becomes clear that there is something more at stake than the effect of binding when Freud relates what he refers to as a "compulsion of destiny"[67] to both the passive experience of compulsive repetition and the active behavior of individuals repeatedly finding themselves in the same situations. Subjects tormented by the repetition of the same unpleasurable situations find themselves possessed or haunted, Freud reports, by something like a " 'daemonic' power."[68] While Freud sets out to investigate how this daemonic face of compulsive repetition can or cannot accord with the pleasure principle, he seems bewildered by the lack of astonishment that we have for repeated voluntary behaviors: "This 'perpetual recurrence of the same thing' causes us no astonishment when it relates to *active* behavior on the part of the person concerned and when we can discern in him an essential character trait, which always remains the same and which is compelled to find expression in a repetition of the same experiences."[69] If the concept of destiny or fate still bore an air of divine ordainment, Freud demystifies it. Destiny is intelligible on a rational basis; it is akin to what we define as "character" or "character-traits," nothing more than a symptom or an effect of a materialistic obstinacy at the core of organic life.[70]

This material transmutation of the core of character and of fate also has implications for our understanding of the ontological significance of habit. Here *Analysis Terminable and Interminable* is relevant, especially where Freud reflects on the constitutive significance of psychological defense mechanisms in the formation of the ego:

> The adult's ego, with its increased strength, continues to defend itself against dangers which no longer exist in reality; indeed, it finds itself compelled to seek out those situations in reality which can serve as an approximate substitute for the original danger, so as to be able to justify, in relation to them, its maintaining its habitual modes of reaction.[71]

From this passage, we learn that the tendency to repeat is not necessarily bound to a particular object. Instead, the object appears to be merely a contingent placeholder for the fixation itself. Hence, any object will do as long as it satisfies the subject's need to play out his original obsessive character. This indifference toward the object pole, and the tendency to seek out situations in which the repetition can be played out marks the same distancing between repetition and what is repeated that Malabou perceived as a non-binding repetition, but rather than manifesting the power of transformation it here announces an original resistance to change throwing the subject back upon itself, regardless of pleasure. From this perspective, habit appears in a new light. Above, we articulated the tendency of habit to "mortify." James as we saw famously described habit as the "fly-wheel" of society, fixating people's behavior and enslaving them in social structures. What Freud suggests in this passage is that the ego appears to seek out situations that can justify his habits. Whereas habit is often seen as a phenomenon which, for better or worse, progressively makes us cling to certain objects and manners, Freud appears to suggest that we cling or adhere to places, things, and people in order to protect our habits. This can be interpreted in two ways: either the compulsion of habit is a direct manifestation of a sort of entropy at the core of being, in which case the fact that we cling to places where our habits can unfold manifests the gravitational pull of this inherent inertia, or habit serves a veil or compromise that conceals the fact that we do compulsively repeat.[72] While it is beyond present analysis to conclude on this matter, it becomes clear that the conceptualization of the compulsion to repeat offered by Freud challenges the traditional conceptions of the so-called force of habit. Being neither a progressive fossilization or sedimentation of life or spirit, nor a dynamic exaltation of being, the compulsion of habit could in Freudian terms be seen as the manifestation of an original resistance to change maintaining the individuals circuitous path to death.

CONCLUSION

For Hume, the propensity of habit constituted an epistemological threshold. Ravaisson understood the propensity to develop habits as an exaltation of spirit, the metaphysical signpost of his spiritualist philosophy. I have tried to account for the propensity of habit from a materialistic perspective. The crux of this attempt is the intention to account for the futile perseverance of habit, which seems to offer a challenge to the positive conception of habit as a phenomenon of adaptive plasticity. Confronted on the point of compulsive repetition, the encounter between Freud and Malabou disclosed a fundamental ambiguity in the *pharmakological* motor of habit. Whereas for Malabou, the compulsion to repeat is a plastic force, the ingenious and ultimately creative work of destructive plasticity, Freud to the contrary saw the blind perseverance of repetition as the manifestation of an inherent organic elasticity, casting the individual back upon an earlier form. The implications of these views differ radically. Hence, understanding habit and the propensity to repeat on the basis of Malabou's threefold concept of plasticity implies that habit is a phenomenon of change—even the stubborn propensity to repeat must be understood as a transformative power. As the echo of an untouchable deadweight within life, the propensity of habit to renew the same actions would from a Freudian perspective not be an acquired tendency, but an immanent persistence out of which the subject's character is spun as a repetition of its own inert obstinacy. Whether the tendency to repeat manifested in the phenomenon of habit should be seen as a transformative ontological potential, or on the contrary as the expression of an inherent inertia resisting change, is in the end a question of metaphysics, but it also concerns the ontological role we attribute habit. Deleuze famously stated: "we are habits, nothing but habits—the habit of saying 'I'."[73] Rather than concluding univocally in favor of one these interpretations of the propensity of habit,[74] which would require that we also come to a definition of the self, I wish to return to Malabou's challenge to *us* as thinkers confronted with the pharmakon habit, to emphasize the necessity to question the conceptual framework, or the *habitués* of thought, with which we seek to understand the nature of habit. It might well be, that with the propensity of habit we are confronted with an elasticity, a daemonic and constitutive return, that challenges the rule of plasticity, in both its negative and positive dimensions.

NOTES

1. This research is funded by the Research Foundation – Flanders (FWO).

2. Catherine Malabou, *The Future of Hegel: Plasticity, Temporality and Dialectic*, trans. Lisabeth During (London: Routledge, 2005), 190.

3. Catherine Malabou, "Addiction and Grace," in Félix Ravaisson, *Of Habit*, trans. Clare Carlisle and Mark Sinclair (London: Continuum, 2008), xix; *The Future of Hegel*, 55–76; Slavoj Žižek, "Discipline between Two Freedoms—Madness and Habit in German Idealism," in *Mythology, Madness, and Laughter: Subjectivity in German Idealism*, ed. Slavoj Žižek and Markus Gabriel (London: Continuum, 2009), 100; *Less Than Nothing: Hegel and the Shadow of Dialectical Materialism* (New York: Verso, 2012), 341–44. Malabou and Žižek both uncover Hegel as the thinker of the pharmakon-habit. We may recall the famous passage in the "Anthropology" where Hegel writes that it is "the habit of living which brings on death, and which, when completely abstract, constitutes death itself" (G. W. F. Hegel, *Hegel's Philosophy of Mind*, trans W. Wallace and A. V. Miller [Oxford: Clarendon Press Hegel, 1971], 143). To live, we must attain the habit of living. But it is also from this habit that we die. This does not imply that the work of habit pertains merely to the rudimentary functioning of the human being, habit mediates the universal and the singular in the individual, molding the body as a tool of the soul.

4. G. W. F. Hegel, *Philosophy of Mind*, trans. W. Wallace and A. V. Miller (Oxford: Clarendon Press Hegel, 1971), 143.

5. Thomas Reid, *Essays on the Active Powers of the Human Mind* (London: MIT Press, 1969), 114–17.

6. Clare Carlisle, *On Habit* (New York: Routledge, 2014), 9, 94–95; Malabou, "Addiction and Grace," vii–viii.

7. See Harry G. Levine, "The Discovery of Addiction: Changing Conceptions of Habitual Drunkenness in America," *Journal of Studies on Alcohol* 39 (1978): 143–74.

8. See Elizabeth Grosz, "Habit Today: Ravaisson, Bergson, Deleuze and Us," *Body & Society* 19, no. 2 and 3 (May 2013): 217–39; Andrew Lapworth, "Habit, Art, and the Plasticity of the Subject: The Ontogenetic Shock of the Bioart Encounter," *Cultural Geographies* 22, no. 1 (2015): 85–102. For the relation between

neuroplasticity and habit, see Ann M. Graybiel, "Habits, Rituals, and the Evaluative Brain," *Annual Review of Neuroscience* 31 (2008): 359–87. Clare Carlisle nuances the question concerning the relation between plasticity (and neuroplasticity) and habit, when she writes: "While contemporary accounts of the brain's plasticity help us to understand the processes of habit formation, philosophical reflection on habit helps us to understand the significance of plasticity" (Carlisle, *On Habit*, 22.)

9. William James, *The Principles of Psychology*, vol. 1 (London: Macmillan and Co., 1891), 105.

10. In a recent article, Elizabeth Grosz defines habit as the instantiation of "open-ended plasticity." This implies first that habit is a temporally open phenomenon, meaning that the future is not contained and determined by present disposition of habit, but merely anticipated, and secondly that habit "provide the ability to change one's tendencies, to reorient one's actions to address the new, and to be able to experience the unexpected" (Grosz, "Habit Today," 217, 221).

11. Catherine Malabou, "Plasticity and Elasticity in Freud's *Beyond the Pleasure Principle*," *Diacritics* 37, no. 4 (2007): 80.

12. Malabou, "Addiction and Grace," ix.

13. Malabou, *The Future of Hegel*, 74.

14. James, *The Principles of Psychology*, 109, 121.

15. Malabou, *The Future of Hegel*, 190. For an interesting account on the idea of habit as fossilization, see Mark Sinclair, "Is Habit the 'Fossilized Residue of Spiritual Activity? Ravaisson, Bergson, Merleau-Ponty," *Journal of the British Society for Phenomenology* 42, no. 1 (January 2011): 33–52.

16. David Hume, *An Enquiry Concerning Human Understanding*, ed. Tom L. Beauchamp (Oxford: Oxford University Press, 1999), 122.

17. Hume, 121.

18. Gilles Deleuze, *Difference and Repetition*, trans. Paul Patton (London: Continuum, 1994), 94. Deleuze writes: "Given that repetition disappears even as it occurs, how can we say 'the second,' 'the third' and 'it is the same?' " (90). What reoccurs as separate instances in repetition has no "in-itself," thus the condition of possibility of repetition is in the difference between the instances contracted by the being that contemplates the repetition.

19. Hume, *An Enquiry*, 121.

20. Mark Sinclair, "Ravaisson and the Force of Habit," *Journal of the History of Philosophy* 49, no. 1 (2011): 65–67.

21. Félix Ravaisson, *Of Habit*, trans. Clare Carlisle and Mark Sinclair (London: Continuum, 2008), 53.

22. Ibid., 51.

23. Ibid., 55, 67, 77.

24. Ibid., 63. This exaltation of spontaneity, perpetually propelling being to repeat itself in its reoccurrence, might be what Deleuze refers to when he writes: "There is indeed an activity of reproduction which takes as its object the difference to be bound; but there is more profoundly a passion of repetition, from which emerges a new difference" (Deleuze, *Difference and Repetition*, 120).

25. Catherine Malabou, *The New Wounded*, trans. Steven Miller (New York: Fordham University Press, 2012), 196; Malabou, "Plasticity and Elasticity," 83, 85.
26. Malabou, *The New Wounded*, 195–98.
27. Ibid., 197.
28. Sigmund Freud, "Beyond the Pleasure Principle," in *The Standard Edition of the Complete Psychological Works of Sigmund Freud*, vol. XVIII, trans. J. Strachey with A. Freud (London: Vintage Press, 2001), 26–27, 29.
29. Ibid., 29–30.
30. Ibid., 30–32.
31. Ibid., 32.
32. In the case of the child's repetitive "fort-da" game, Freud suggests there is a will to mastery underlying the unpleasant repetition. In this form Freud suggests that the compulsive repetition appears to work independent of the pleasure principle, however not in opposition to the tendency of this principle (Freud, 16–17).
33. Malabou, *The New Wounded*, 197.
34. Ibid., 197; Malabou, "Plasticity and Elasticity," 85.
35. Malabou, *The New Wounded*, 197–98.
36. Ibid., 199.
37. Ibid., 198.
38. Ibid.
39. Ibid.
40. Malabou, *The New Wounded*, 180; Gilles Deleuze, *Spinoza: Practical Philosophy*, trans. Robert Hurley (San Francisco, CA: City Lights Books, 1988), 17–18.
41. Malabou, *The New Wounded*, 200.
42. Ibid., 199–200.
43. Ibid., 198.
44. Catherine Malabou, *Ontology of the Accident*, trans. Carolyn Shread (Cambridge: Polity Press, 2012), 22.
45. Deleuze, *Difference and Repetition*, 94, 100, 121. Here it must also be added that binding for Deleuze's has nothing to do with mastering. There is no preexisting agency underlying the passive synthesis serving as the contractive power of habit.
46. Hume, *An Enquiry*, 122–24.
47. Malabou, *The New Wounded*, 197.
48. Ibid., 199.
49. Freud, "Beyond the Pleasure Principle," 35.
50. Ibid., 36.
51. Ibid.
52. Ibid. Original italics.
53. Malabou, "Plasticity and Elasticity," 82; *The New Wounded*, 177; *Ontology of the Accident*, 36. It must be added that there is more than one definition of "elasticity." The definition Freud uses, and the one Malabou refers to here, stems from physics and is defined as the ability of a body to resist external influence and its capacity to return to its initial form when the external excitation is removed. If we take Henri Bergson as an example, we see that "elasticity" has a very different definition. For

Bergson elasticity is the opposite of automatism and obstinacy, and thus not understood as the character of compulsive repetition. Elasticity on the other hand denotes the living being's capacity to change and adapt to the continuous unfolding of life. It is the characteristic of the soul that molds the body. Elasticity is what enables the individual to resonate with the world, the social as well as the natural (*Laughter: An Essay on the Meaning of the Comic*, trans. C. Brereton and F. Rothwell [New York: Dover Publications, 2005], 9–10, 14).

54. Malabou, "Plasticity and Elasticity," 80–81, 83; *The New Wounded*, 173–78.

55. Malabou, "Plasticity and Elasticity," 82.

56. Malabou, "Plasticity and Elasticity," 83, 85; *The New Wounded*, 196–98. Malabou also critiques Deleuze for his usage of the term, when he refers to the "elasticity" of the *conatus* (Malabou, *Ontology of the Accident*, 36).

57. Malabou, *The New Wounded*, 15.

58. Sigmund Freud, "Analysis Terminable and Interminable," in *The Standard Edition of the Complete Psychological Works of Sigmund Freud*, vol. XXIII, trans. J. Strachey with A. Freud (London: Vintage Press, 2001), 219.

59. Freud, "Beyond the Pleasure Principle," 38.

60. Ibid., 38.

61. In this text I do not differentiate between "acceptance" and "receptivity."

62. Freud, "Analysis Terminable and Interminable," 242.

63. Ibid.

64. Freud, "Beyond the Pleasure Principle," 39.

65. Ibid., 44.

66. Ibid., 23.

67. Ibid.

68. Ibid., 21, 35, 36.

69. Ibid., 22.

70. Ibid., 22–23. Bergson also refers to something like a materialistic obstinacy within the individual, but contrary to Freud he defines this as "inelasticity." As the sign of "separatist tendencies" this is at the basis of the comic, something to be corrected by laughter as a means of social correction (Bergson, *Laughter*, 10).

71. Freud, "Analysis Terminable and Interminable," 238.

72. Beneath these two possible readings we are confronted with the question concerning primary repression, which Deleuze exposed as the split question: Do we repeat because we repress, or, do we repress because we repeat (Deleuze, *Difference and Repetition*, 130)? Whereas Deleuze clearly decides for the latter, it is still debated whether or not Freud's theory of the drives can afford such a reading. Here, it should not surpass us that Deleuze in *Difference and Repetition* offers another reading of compulsive repetition, along with this theory of habit as phenomenon of passive synthesis.

73. Gilles Deleuze, *Empiricism and Subjectivity*, trans. C. V. Boundas (New York: Columbia University Press, 1991), x.

74. It must necessarily be added that Deleuze offers a third position in *Difference and Repetition*, which opens up on the basis of a critical assessment of Freudian

psychoanalysis. Consequently, the next step in a philosophical questioning of the propensity of habit must include Deleuze's critique of Freud's conception of compulsive repetition and a discussion of Malabou's critique of Deleuze.

BIBLIOGRAPHY

Bergson, Henri. *Laughter: An Essay on the Meaning of the Comic*. Translated by Cloudesley Brereton and Fred Rothwell. New York: Dover Publications, 2005.

Carlisle, Clare. *On Habit*. New York: Routledge, 2014.

Deleuze, Gilles. *Empiricism and Subjectivity*. Translated by Constantin V. Boundas. New York: Columbia University Press, 1991.

———. *Difference and Repetition*. Translated by Paul Patton. London: Continuum, 1994.

———. *Spinoza: Practical Philosophy*. Translated by Robert Hurley. San Francisco, CA: City Lights Books, 1988.

Freud, Sigmund. "Beyond the Pleasure Principle." In *The Standard Edition of the Complete Psychological Works of Sigmund Freud*, vol. XVIII, translated by James Strachey with Anna Freud. London: Vintage Press, 2001.

———. "Analysis Terminable and Interminable." In *The Standard Edition of the Complete Psychological Works of Sigmund Freud*, vol. XXIII, translated by James Strachey with Anna Freud. London: Vintage Press, 2001.

Graybiel, Ann M. "Habits, Rituals, and the Evaluative Brain." *Annual Review of Neuroscience* 31 (2008): 359–87.

Grosz, Elizabeth. 2013. "Habit Today: Ravaisson, Bergson, Deleuze and Us." *Body & Society* 19, no. 2 and 3 (May 2013): 217–39.

Hegel, Georg Wilhelm Friedrich. *Philosophy of Mind*. Translated by William Wallace and A. V. Miller. Oxford: Clarendon Press, 1971.

Hume, David. *An Enquiry Concerning Human Understanding*. Edited by Tom L. Beauchamp. Oxford: Oxford University Press, 1999.

James, William. *The Principles of Psychology*, vol. 1. London: Macmillan and Co., 1891.

Lapworth, Andrew. "Habit, Art, and the Plasticity of the Subject: The Ontogenetic Shock of the Bioart Encounter." *Cultural Geographies* 22, no. 1 (2015): 85–102.

Levine, Harry G. "The Discovery of Addiction: Changing Conceptions of Habitual Drunkenness in America." *Journal of Studies on Alcohol* 15 (1978): 493–506.

Malabou, Catherine. *The Future of Hegel: Plasticity, Temporality and Dialectic*. Translated by Lisabeth During. London: Routledge, 2005.

———. *The New Wounded: From Neurosis to Brain Damage*. Translated by Steven Miller. New York: Fordham University Press, 2012.

———. *Ontology of the Accident: An Essay on Destructive Plasticity*. Translated by Carolyn Shread. Cambridge: Polity Press, 2012.

———. "Plasticity and Elasticity in Freud's *Beyond the Pleasure Principle*." *Diacritics* 37, no. 4 (2007): 78–85.

———. "Addiction and Grace: Preface to Félix Ravaisson's *Of Habit*." In *Of Habit*, translated by Clare Carlisle and Mark Sinclair. London: Continuum, 2008.

Ravaisson, Félix. *Of Habit*. Translated by Clare Carlisle and Mark Sinclair. London: Continuum, 2008.

Reid, Thomas. *Essays on the Active Powers of the Human Mind*. London: MIT Press, 1969.

Sinclair, Mark. "Ravaisson and the Force of Habit." *Journal of the History of Philosophy* 49, no. 1 (2011): 33–52.

———. "Is Habit the Fossilized Residue of Spiritual Activity? Ravaisson, Bergson, Merleau-Ponty." *Journal of the British Society for Phenomenology* 42, no. 1 (2011): 33–52.

Žižek, Slavoj. "Discipline between Two Freedoms—Madness and Habit in German Idealism." In *Mythology, Madness, and Laughter: Subjectivity in German Idealism*, by Slavoj Žižek and Markus Gabriel. London: Continuum, 2009.

———. *Less Than Nothing: Hegel and the Shadow of Dialectical Materialism*. New York: Verso, 2012.

Chapter 5

Attached to Detachment

A Materialist Indifference in Catherine Malabou[1]

Cristóbal Durán

Do we still have to presuppose a gap between the structural and the material in order to render the material meaningful? Do we have to transcend the empirical organization of the real in order to produce a theory of the real? Or shouldn't we on the contrary consider such gestures as sovereign acts which reinscribe, just as kingship, the excess at the heart of meaning?[2]

—Catherine Malabou

A MATERIALISM OF TRANSITION

In several recent works, Catherine Malabou proposes all kind of heterogeneous ideas—some more, some less explicit—that help further the discussion on the reach and future of materialism. We could say that materialism is at the heart of her theoretical concerns, and we could even take the more extreme view that materialism is what marks or defines the style of interrogation that makes these concerns intelligible. If, as Gérard Granel observed about Marx, "the materialist thinker is precisely someone who thinks, not of the material, but of the formal,"[3] we may begin to assume that material's transformability is possibly what gives materialism its broader meaning. This means that the material is not defined by its limitations—for instance, in a simple opposition between matter and form—but rather as form's *ligne de fuite*, a sort of immanent labor of deformation.

As such, the *-ism* of materialism doesn't refer to an exacerbated material trait, which presumably applies to matter in its unformed state. The meaning of the material is excessive in that it exceeds the lapse and pushes beyond the boundary that distinguishes the structural and the material. Instead of simply rejecting the idea of a hollow form that can be filled with material content, Malabou proposes ideas that allow us to dislocate the pure fixation and identification of a structure. In *Plasticity at the Dusk of Writing*,[4] Malabou attempts to show that the idea of form inherent in plasticity needs to be developed without recourse to the trace or archi-trace. According to Malabou's analysis, in Levinas and Derrida the trace appears to be divorced from form, it resists the formations that dictate the movement of presence. The trace is separate from its own movement.

Beyond form—especially when we consider Levinas—the trace is transcendent; it can be thought of as a field of alterity that doesn't require the presence of forms. That is, it surpasses form and resists the metamorphic abilities of convertibility. In this way, the trace is safeguarded against convertibility. Likewise, Malabou understands the *Différance* as a resistance to form. This is because the concept of form retains and contains the intensity of its genesis and production, and is inevitably marked by delay. If, as Derrida states, the form is presence itself, then any metamorphic resource only serves to outdo presence, leaving the form firmly situated within the conceptual realm of presence. It's from within this realm that Malabou attacks the transcendence implied by separation, as a kind of fetishism of the form—that is, the idea that form is inseparable from trace—that ultimately serves to dematerialize contemporary philosophy.[5]

Rather, Malabou argues that there is no way to transcend the form, there is no transcendence of the form, and actually, form is "the metamorphizable but immovable barrier of thought."[6] Arguably, by focusing on the metamorphizable and immovable barrier, we can discern a concept of form that is permeated by materialism—this is nothing less than the movement of absolute convertibility that is impossible to separate from form. Changing from one form to the next, we begin to see an "excess of form in regard to presence";[7] the conventional concept of form is superseded, in need of being reworked in order to be understood as a process of transformation. What Malabou refers to as "New Materialism" refers to an alterity that is no longer a limit to interiority. To quote Malabou: "Alterity can be thought without the aid of transcendence, if it is true that there is nothing outside, nothing beyond—outside the economy and outside the exchangeability or mutability of Being—then *there is no inconvertibility. Absolute convertibility, the migratory and metamorphic resource of alterity, is the rule. Absolute exchangeability is the structure*."[8]

"New Materialism" points to the plasticity beyond destruction and deconstruction, or in other words, to the fact that it is impossible to draw a line between form and that which resists form. The material of this materialism will always remain plastic: it retains an imprint and thereby resists endless and absolute polymorphism. If the form is fixed and becomes an inconvertible trace, then this is arguably the result of form fetishism. Against form, the otherness is fetishized. "The assertion of inconvertibility lies, for Marx, at the heart of fetishism. On the face of it, the fetish always occurs outside the operation of exchange, outside the market."[9] This is not only a nonrejection of forms but a reinforcement of them; that is, it draws attention to forms in order to show their nonabsolution and nonseparateness. Arguably, this is the first way in which plasticity puts us on the path toward a materialism defined by the metamorphosis of immanence, that is, the unprecedented movement that shapes formation and deformation.

The model for this materialism is explicitly described in *What Should We Do with Our Brain?* This text makes the argument for a "Reasonable Materialism," one that attempts to show that the psychic is a transformation or a metamorphosis of the neuronal, and that it doesn't involve a fusion of these two concepts any more than it does a clean separation between their domains. The book attempts to conceive "the transition from the undifferentiated to the possibility of a transdifferentiation of self."[10] One would have to accept that, as Malabou states, "the necessary mediation of the idealization of self—that the position of neuronal materialism . . . should elaborate a central idea, or theory, of the transition."[11] This idealization allows us to go back and discern a certain materialism of transition. Said materialism does not attempt to fuse these two domains, nor maintain them separate from one another. Rather, this materialism consists of a transition in which the natural contradicts itself.[12] The immovable, yet metamorphizable, barrier is defined as transition and transdifferentiation: this transition doesn't take place between one state and another; rather it's a formation of differences. What we will attempt to show in the paragraphs that follow is that the immanence we are beginning to discern rests on a hinge, on a lapse in which the formation of immanence encompasses neither differences nor clearly defined distances. As such, it can neither separate itself from itself nor create a transcendental horizon from which to program its own future.

In a recent text entitled "Will Sovereignty Ever Be Deconstructed?," Malabou tries to show that life's formative processes cannot be attributed to a transcendental resource. Malabou challenges the concept of life found in Foucault or Agamben, and she argues that the entire philosophical tradition against biology ultimately generates opposition between two types of life. Moreover, she argues that something like a symbolic life gives a rigid, transcendental meaning to the symbolic: "It is as if we still needed to affirm the existence of a beyond or an outside of the real to confer meaning to reality, as if a prior structure, necessarily nonmaterial, was requested to give sense to materiality itself."[13] This separation ultimately means that the material is animated from without, thus preventing us from seeing materiality through the lens of its own symbolic production and creating a "transcendental symbolic economy."[14] As Malabou warns us, the main problem with this is that it confines the biological to a mere realm of predetermination: the biological becomes "a simple raw material for political use."[15]

So, how should we understand this transdifferentiation or metamorphosis that arguably defines materialism? Let us consider *in extenso* the following definition provided by Malabou:

> Materialism is a name for the nontranscendental status of form in general. Matter is what forms itself in producing the conditions of possibility of this formation itself. Any transcendental instance necessarily finds itself in a position of exteriority in relation to that which it organizes. By its nature, the condition of possibility is other than what it makes possible. Materialism affirms the opposite: the absence of any outside of the process of formation. Matter's self-formation and self-information is then systematically nontranscendental.[16]

Matter generates its own possibilities and, in doing so, forms a posterity that it cannot see coming; it offers up its possibilities without knowing what to expect and without full knowledge of the direction in which it is headed. As such, there is no transcendental resource that determines formation, and formation must free itself from itself in order to give itself a future. This is precisely one of Malabou's more pointed lines of attack in her previous interpretation of the emotional brain. Malabou finds that there's a "virtual coldness" or "power of annihilation" in the fate of organic brain damage, and that this coldness reveals "a law of being that always appears to be on the point of abandoning itself, escaping."[17] The damage reveals a condition that seems to be rooted in cerebral self-affection: the first relationship that cerebral life has with itself, and that ultimately serves as the foundation for psychic life, is life's attachment to itself, one that, notwithstanding, knows nothing of itself. Strictly speaking, there is no nexus between affection and selfhood, or, to put it differently, this is a basic attachment with absolutely no knowledge of (dis)attachment. The very heart of subjectivity rests on a strictly material detachment, an interval in which subjectivity is formed without seeing what is coming. The lapse inscribed by the cerebral accident is entirely material. This, in the sense that this lapse forms a posterity without succession and is unaware of the sequence of events that preceded and produced it. Without any programmed or foreseen future, the only tie that binds the Self to itself is given by an "affective economy that solicits itself *without seeing itself*."[18] Indifference, far from being excluded from the production of subjectivity, permeates the immanence wherein subjectivity is defined by its attachment to itself. This economy is what Malabou refers to as the "Material," and it contradicts the three orders described by Lacan: the organization and regulation of the brain makes it possible to conceive a radical lack of struggle for recognition. This implies a difference between a normal brain and a damaged brain, and it follows that "if the 'normal' brain never affects itself without meeting up with itself in the mirage of its own separation, the damaged brain has no chance, *a fortiori*, of being present to its own fragmentation or its own wounding."[19]

This materiality—and this is precisely what interests us—indicates that there is a moment in which the separation cannot foresee the result of a damaged cerebral connection. This results in a strictly material lapse, where the brain can't anticipate its own damage.[20] Here, materiality must conceive an other, that is not the other of alienation or specularity. This materialist logic maintains the movement of formation and deformation in neuronal connections. This is a metamorphosis of a life that dies from affective indifference, and as such, it reveals a kind of impossible separation at the very center of life.[21] Be this as it may, it's important to realize that this indifference doesn't produce homogeneity and it doesn't require us to conceive a materialism that is identical to itself. It's defined by neither mere inertia nor subjective spontaneity. The absence of the self to itself[22] continues to be a type of metamorphosis "that is a farewell to being itself. A farewell that is not death, a farewell that occurs within life, just like the indifference of life to life which survival sometimes manifests itself."[23]

It is interesting to note that this relationship of indifference allows us to discern a radicalization of the relationship that defines both subjectivity and the deconstruction of subjectivity. Is it because of this lapse of indifference, that life gives itself the opportunity to always become something other than what it is or "can be" given its programming? Does this indifference encompass the "supplement of indeterminacy" and the "floating entity" that Malabou identifies in Althusser and Darwin? Let us go one step further. Materiality has an inner tendency that allows us to perceive a "self-transformative tendency of life." This possibility defines life as continuous variation, "transforming itself without separating itself from itself."[24] In a strong sense, the formation of life does not know what it is forming, it's a formation that exceeds the very bounds of life.

ATTACHED TO INDIFFERENCE, OR THE DETACHMENT OF A NEW TYPE OF SELF

Let us put it more directly: if the formation of form implies materialism, if the genesis of form is what generally accounts for what that form could not have been, that is, if we insist on the epigenetic factor, which sees the transcendental as the movement of progressive differentiation and emphasizes the unpredictability at the heart of generative becoming,[25] then structure cannot be understood as pure predetermination or programming. As Malabou states, recent developments in neuroscience turn our attention toward the more extreme case in which not only is it possible to show that self-affection depends on the movement inherent in its transformation in order to take form, but, furthermore, we are challenged to conceive a state in which the definition of a subject is no longer merely paired to the affection that the subject is purportedly supported by. More specifically, Malabou refers to cases in which brain damage has led to destructed subjectivity: in the more extreme cases, victims of brain damage seem to prompt the question as to whether or not there can be affects that don't affect "me." "Is the affected subject, then, someone else, the presence of another subject within itself? Or, is it just nobody, the absence of a substantial first person?"[26]

First and foremost, it is important to note that this discussion is not restricted solely to cases of brain damage. Neuroscience allows us to posit the initial affectability of a subject that leads to the "I," and, accordingly, there is no reason to suppose that this "I" knows anything about itself. What has come to be known as cerebral self-affection—a way of signaling that emotions are the most basic way in which the *Self* keeps a firm grasp on itself, a series of unconscious, homeostatic processes that keep an organism alive—is, strictly speaking, a form of biological heteroaffection. In a way, cases of extreme brain damage refer *post facto* to a process that is both rooted in and gives initial rise to the very processes of autoaffection: "The study of the emotional brain precisely challenges the vision of a self-affecting subjectivity in favor of an originary *deserted* subject, a subject that is definitely not present to itself."[27]

In this sense, it is possible to say that neurobiology can explore the absence of the *Self* to itself. Furthermore, neurobiology makes clear that this movement is not just an extreme case in which subjectivity appears to be affected and compromised; rather, it seems to be subjectivity's nature and mode of operation *princeps*. Consequently, and in a very interesting way, Malabou ascribes the heteroaffected (or disaffected: something that remains nonsubjective or nonreferable to a first person) origins of self-affection to an underlying schema of *Wonder*, or *Admiratio*. If *Wonder*, in the work of both Descartes and Spinoza, serves as a keystone for subjectivity in a substantive metaphysical approach to self-affection, the question that comes to light when discussing cases of brain damage is this: What happens when the *Self* is unable to pick up on its *own* affects, when the *Self* is unaware of self-affection?

If this is true, what happens when a brain map—guarantor of our identity—is incapable of assimilating affects into an *autos*? What happens when autoheteroaffection is impaired as a result of brain damage? This produces a new type of figure: unable to subjectively assimilate phenomenon, the passions become isolated from the subject's personal history. A sort of biological deconstruction of subjectivity takes place,[28] the likes of which imply the existence of a heteroaffected subject in their most radical form: it is a becoming-Other of the Other that is no longer consistent with the *Self*. This creates a hetero-heteroaffected subject, who is now completely "disaffected." By disconnecting a subject's brain map from their emotional processes, we are now able to conceive of a *deserted subject*. This desertion is perhaps the best indicator of a subject that is no longer present; far from overlooking this sort of case, it becomes more important when we consider that desertion is a defining aspect of the rudimentary environment within which subjectivity is born. By all accounts, desertion is a test of an affected Other to the point where they are no longer an Other; it signals the broken bonds that would otherwise make autoheteroaffection a guarantor of subjectivity. How to conceive of a nonaffected and nonaffectable subject, to complete or accomplish the deconstitution of traditional subjectivity?[29]

When answering this question, we risk taking the following assumption for granted: we tend to see that very same disaffection as one of the more singular and important cornerstones of Malabou's thinking. Not only is this a result of the type of philosophical thinking that Malabou calls into question, but it also makes us aware of a more fundamental problem belonging to this branch of philosophical thought: there is a strong and unyielding attachment to that from which one hopes to escape. It signals a resistance to detaching ourselves from precisely that whose affect we hope to avoid. In many ways, this issue is visible in Malabou's early reading of Hegel. At the heart of *Aufhebung* lies a "giving up" or a "letting go" that means abandoning its self-chosen position: "just like a hand which opens up and relinquishes what it has been holding. From this comes a relaxing of the tension that had separated thought from its object and preserved their encounter only by maintaining them in this divide."[30] The giving up, or abrogation, that implies a "relaxing of the tension," corresponds to the way in which the "I" relinquishes command over its own preservation and dominion, and depends on a relief or sublation (*relève*) "that frees from a certain type of attachment."[31]

At this point, there is a nearly indiscernible lapse in which subjectivity is lost and then regains its new self; the self regained is different from the self that had previously beheld its own transformation. The Absolute's movement is made possible when the unified retention "relaxes the tension" that would otherwise create a tight bind between a subject and an object. Now, and with greater fluidity, the bind between the two loosens and it is permeated by what it was previously unable to see and/or predict. By surrendering its own mastery, it's not that the "I" is fused with its own thoughts, but that it is able to open itself up to contact that will leave it altered and transformed. The unity, or uniqueness, that exists here is watched over silently and indifferently by a new type *Self*, a *Self* that did not see itself coming. From this point on, it's subjected only to its own detachment, to that which it cannot see coming.

What makes Malabou's thesis radical is that the type of detachment she proposes changes the sort of differences absorbed by the Absolute. Malabou's type of detachment manages to divert the Absolute from its course toward restitution, thus renewing it time and time again, all of this in the absence of a sequence that would permit a subject to recognize themselves at some future point in time. Following Judith Butler's footsteps, Malabou understands the relationship between domination/servitude (from the French reception on Hegel's master-slave dialectic) as part of the schema of the subjection and difference between Attachment and Detachment. Malabou wonders if it's possible to make the sort of statement that, according to Butler, is made by the bondsman: "You be my body for me, but do not let me know that the body you are is my body."[32] The question being asked here is whether it's possible to propose a definitive disappropriation of one's body as well as a definitive appropriation of another's body. Yet, according to Malabou, the body (which seems to be subject to itself) is actually shaped by something outside of itself, that is, it finds its form via the body of an Other. As a result, the Other's body ceases to "belong" solely to the Other. For Malabou, avoiding attachment altogether is impossible. Thus, resistance to it depends, not on locating some defect in the integrity of the concept, but on the fact that attachment itself cannot be absolute. At the same time, the idea of an absolute detachment supports the notion of a transcendent body, or consciousness, which is able to exercise total control over another.

When self-affection is put in check (which presupposes the length of the fold needed for the *Self* to retrieve and recover itself), heteroaffection likewise becomes disputable. This is due, in the first instance, to the fact that the affect one has on an Other is indicative of an attachment to attempting to be outside of oneself; in other words, it is an attempt at resisting the changes wrought by time. "This implies that detachment always entails some attachment. Indeed, the very act of claiming that absolute detachment is necessary reveals an attachment to it; otherwise, why would it have to be claimed?"[33] Put differently: given that neither the body nor consciousness are capable of affecting or emotionally moving themselves, they remain alien to themselves. The result is that not only does the "I" not affect itself, but the Other is always outside-of-itself. This, first and foremost, is an inclination toward Otherness that is nonetheless always attached to itself. This is true even when the distance that separates affection from an *autos* is left up to fate by the blueprint that otherwise lays the tracks for the intersection between auto- and heteroaffection. For better or worse, the *self* of self-affection is tasked with creating the accident that will both cause it to leave itself and permit it to return.

In the very same text, Malabou makes the claim that toward the end of *Phenomenology of Spirit*, "a supreme detachment" is what gives plasticity "its definitive meaning." The Self loses the form of the "I" and it loses the form of its disjunction as well. The form of the "I" explodes and dissolves itself.[34] All of this is to say that it's precisely this sort of undoing that, in some sense, defers the process of recovery of spirit. Or, at the very least, it gives it another status and meaning. This absolute substitution (one that is absolved of itself) is the paradox that lies at the heart of the absolute. Detachment, like abandonment or like free detachment, is what consistently prevents a homogeneous recovery of spirit with respect to itself. This ongoing mourning of the present, a lament over its fleetingness and ephemerality, coaxes the *Aufhebung* into play. Without it, plasticity would agree with a purely positive version of dialectics in its movement of negativity and determination.

This abandonment would also not entail a reattachment to the sameness of the Absolute. As Malabou points out, we are focusing here an anonymous self of Spirit, a "new form of self without inwardness, without externality," that "is *subjected* neither to anyone nor to anything."[35] Malabou says that this self is not even attached to itself, suggesting that this absolute detachment is "an impersonal, silent, and so pleasurable ecstatic indifference, which, erasing the limits of what used to be 'our' bodies, *unbinds* us from the chain of continuation."[36] This abrogation transforms the attachment the Absolute maintains with itself, the very attachment that allows the Absolute to continue being what it is.

One can say that this concern leads us to consider that subjectivity is not simply defined in contrast to its previous state and what might be expected from that state. As Michel Serres states, "One always detaches oneself only from comparison."[37] If this recognition by comparison is simply the inability to let oneself go, then what this materialism is trying to propose is an abandonment of self-referentiality, the possibility of conceiving an encounter with that which cannot be compared. Rather than a simple detachment, this indicates a disattachment, one that is so passionate that the attachment never stops disattaching itself from itself, it never stops disassembling. Inseparable from itself, life may be nothing more than the immanent movement of its own material in metamorphosis.

As Malabou states in *The New Wounded*, some separation (or split) is necessary in order to be exposed to future events. To be exposed in general. Because of this, the self-affective model which is put to the test by trauma only becomes visible thanks to disaffection; strictly speaking, the formative accident is impossible for the subjective self to foresee, given that it is bound by a limited set of points of reference. As such, a disturbance in cerebral self-affection wouldn't necessarily call for an end to psychic life (it's as if the brain, without affection, were to survive its own "death"; in a sense, it is disaffected). Furthermore, what's at stake here is a way of understanding affectability that doesn't equate it with a recovery or recuperation: "To a certain extent it is a matter *of* the other of the self *in* the self; but, here, *it is the self, and not the other, who never lets itself be encountered* when traumatized. It is the self who is lacking, without specular recuperation."[38]

In the brain, the affect is neither detached from itself, nor does it form a unified whole. "Indeed, the 'self,' at its very core, is not gathered."[39] In this sense, disaffection reveals the affection the brain has to itself; it is a sort of conatus that, unacknowledged, perseveres. If we return to the lexicon used in *The Future of Hegel*, what's going on here is a true "economy of loss" that makes it necessary to rethink the logic behind conservation. "Maintenance always demands distress (*perdition*): without that, how can there be a threat of loss (*déperdition*) to counteract?"[40] Loss can be addressed once the sequence is cast off, and the only way to continue is by not seeing what's coming. Here, loss has no recourse, nor is it subject to appeal; rather, it creates a new subjectivity that no longer recognizes the preceding one.

If the only possible experience of cerebral autoaffection is one produced by a singular type of suffering (the result of disaffection), then we are forced to acknowledge that the brain, unable to see itself, resigns itself to its own lack of self-expectations. The brain abandons itself so as not to see itself during this lapse. Is this the sign of a considerable differential within the very idea of plasticity, in which there is a very excess of future in the future? That disaffection is unable to see itself and that it consigns the random fate of accidents to some unprecedented future goes hand in hand with the free abandonment and abrogation that allows the Absolute to unbind itself (here, subjugation should not be mistaken for attachment). Perhaps, this occurs when it is no longer able to see itself coming, thus making it necessary to take abrogation into account in the midst of sublation. "If sublation can sublate itself, it is precisely because in the end it *sees itself* and can therefore *miss* (itself)."[41]

Here, losing oneself doesn't equate to death. The negative nature of damage, of sheer fulmination, that brings about disaffection marks a new turn in the deconstruction of subjectivity. A sort of "a law of being that always appears to be on the point of abandoning itself, escaping," and prescribes "farewell of being to himself . . . just like the indifference of life to life by which survival sometimes manifests itself."[42] In this sense, disaffection signals the fact that something unpredictable has been created, something unable to see itself coming. This is not sublation *per se*, given that there is no affection that determines the fate of future affections. This is the lapse in which, rather than be at odds with one another, both the auto and the hetero become entwined. This new psyche can no longer be taken to be a replica or substitution of the preceding one; rather, it is a metamorphosis that is unable to make itself present at the moment of transformation. Indeed, it's even possible for Malabou to say, "It is no longer the same subject who anticipates himself and sees himself die."[43] What disaffection reveals is the existence of a detour and an inability to recover negativity, all because disaffection isn't bound to a set of preliminary factors that account for its possibility. Devoid of both transcendental and transcendence, someone who experiences surprise is no longer capable of being affected by that surprise. Does this allow us to conceive of some alternate future?

Strangely, and in an extremely unique way, disaffection marks the point at which, within the normal course of events, time seems to stop. In a profound and radical way, a new time bursts forth. This is also the force of detachment which makes it possible for life to adhere to its own preservation, even in the absence of an actor who is able to feel it or assimilate it into themselves. If plasticity is the heteroaffective structure of subjectivity, then isn't this impersonality, a sort of synthesis devoid of affection (disaffected, as it were), what makes it possibility to maintain a mutual separation between anticipation of the Same and anticipation of the unexpected future? We can take this argument further. No doubt, it's not a question of showing that disaffection has some underlying structure that precedes plasticity. To do so would mean carrying out a general and far-reaching analysis of modern philosophy through the lens of affection. Above all, I think it's interesting to show how a small, seemingly regional trait or motive, when expanded upon, can help us comprehend the sort of problems Malabou raises concerning the philosophical tradition. It is perfectly possible to think that disaffection is not just a term that exemplifies the results of the destructive power of plasticity. Rather, it can be understood as a regional and determinate element in the overall concept of plasticity.

Therefore, let us imagine that disaffection creates a time in which there *is no longer* time. Or, more precisely: disaffection exposes a fundamental truth of Hegelianism that echoes throughout the System and that affirms emphatically: "it could have been otherwise, it could not have been otherwise."[44] How to decide? Is it possible to anticipate something when there is nothing to wait for, when there is only the unexpected? What is the time or the fate of that disaffection—a disaffection to which the System is designed to stay true, unable to detach itself?

One can ask whether this destruction, from which recovery disassociates itself from the preceding subjectivity, the two of them rendered incompatible, has a future. No doubt, a future does exist, one that takes on the unique shape of "negative possibility." As such, the effects of an accident have the potential to "introduce into the thickness of succession the undatable bifurcation of destruction" and create a context in which "the repertory of viable forms has reached exhaustion and has nothing else to propose."[45] A sort of "bending" of the link, an asyndeton that serves as a mark of an impossible link,[46] literally creates a type of subjectivity defined by a possibility that is rejected. This rejection, however, is not produced by a preexisting subject. The fundamental question here is how to conceive of a future when there is no preexisting meaning that shapes it. "It bears witness to a power or aptitude of the negative that is neither affirmed nor lacking, a power that *forms*."[47] It's this unforeseeable negative possibility that shape to the future, a future that likewise cannot be predicted. Yet, the fact that this concerns an interruption does not mean that plasticity is left aside. It is vital to remember that plasticity is at once "a *structure of transformation and destruction of presence and the present*."[48] What type of future does this destruction of meaning and experience have in store?

The interruption carries the possibility of something new: it is a possibility that is neither fixed to the interruption, nor is it tied to the past. Devoid of a past, the possibility that arrives suddenly and without warning is never fully realized and never goes away; it endures *a posteriori* to presenting itself. This constitutes what Malabou refers to as "*other possibility*, even if it were an *a posteriori* possibility,"[49] and it endures unfinished and ever pending, even when a subject is unable to feel themselves. A possible perpetually pending open to the vertigo of a completely different origin. Events could always have occurred in different ways; it's a continuous question as to how the future could have been. Thus, if negative possibility can insert itself into history without actually being contingent to it, this is because the future doesn't fully follow from the present. There is no corroborated present: disaffection in the present makes it impossible to connect the present with any past affection that might conceivably lend meaning to the future. Thus, is it possible to say that the affect has yet to be created, and that furthermore, it is incompatible with a present that depends on a past-present? Perhaps disaffection is not so much the death of affect, but is instead a sign of an even more extreme attachment capable of engendering affects devoid of heritage, affects that still have to see it coming.

NOTES

1. This chapter is a result of Fondecyt Project 11150732, "El dinamismo singular de la relación: elementos para una reconstrucción de la ontología relacional a partir de una teoría de las multiplicidades (Simondon, Deleuze)," funded by the National Commission for Scientific and Technological Research (CONICYT).

2. Catherine Malabou, "Will Sovereignty Ever Be Deconstructed?," in *Plastic Materialities: Politics, Legality, and Metamorphosis in the Work of Catherine Malabou*, ed. Brenna Bhandar and Jonathan Goldberg-Hiller (Durham, NC: Duke University Press, 2015), 45.

3. Gérard Granel, "Le concept de forme dans 'Das Kapital,'" in *Granel, l'éclat, le combat, l'ouvert*, ed. Jean-Luc Nancy and Élisabeth Rigal (Paris: Éditions Belin, 2001), 28.

4. Catherine Malabou, *Plasticity at the Dusk of Writing: Dialectic, Destruction, Deconstruction* (New York: Columbia University Press, 2010), 44–50.

5. Ibid., 45.

6. Ibid., 49.
7. Ibid., 49.
8. Ibid., 47.
9. Ibid., 77.
10. Catherine Malabou, *What Should We Do with Our Brain?* (New York: Fordham University Press, 2008), 70.
11. Ibid., 69.
12. Ibid., 82.
13. Malabou, "Will Sovereignty Ever Be Deconstructed?," 40–41.
14. Ibid., 43.
15. Ibid., 44.
16. Malabou, "Whither Materialism? Althusser/Darwin," in *Plastic Materialities: Politics, Legality, and Metamorphosis in the Work of Catherine Malabou*, 48.
17. Malabou, *Ontology of the Accident: An Essay on Destructive Plasticity* (Cambridge: Polity Books, 2012), 37
18. Malabou, *The New Wounded: From Neurosis to Brain Damage* (New York: Fordham University Press, 2012), 140.
19. Ibid., 140.
20. Ibid., 141.
21. Ibid., 212.
22. Malabou, "Go Wonder: Subjectivity and Affects in Neurobiological Times," ed. Adrian Johnston and Catherine Malabou, *Self and Emotional Life: Philosophy, Psychoanalysis, and Neuroscience* (New York: Columbia University Press, 2013), 72.
23. Malabou, *Ontology of the Accident*, 37–38.
24. Malabou, "Will Sovereignty Ever Be Deconstructed?," 45.
25. Malabou, 2016, 64.
26. Malabou, "Go Wonder: Subjectivity and Affects in Neurobiological Times," 8.
27. Ibid., 4.
28. Ibid., 58.
29. Ibid., 60.
30. Malabou, *The Future of Hegel: Plasticity, Temporality and Dialectic* (London: Routledge, 2004), 157.
31. Ibid., 156.
32. Judith Butler, *The Psychic Life of Power: Theories in Subjection* (Palo Alto, CA: Stanford University Press, 1997), 35.
33. Malabou, "Unbind Me," in *A Companion to Hegel*, ed. Michael Baur and Stephen Houlgate (Oxford: Wiley-Blackwell, 2011), 614.
34. Ibid., 624.
35. Ibid., 624.
36. Ibid., 624.
37. Michel Serres, *Detachment* (Athens: Ohio University Press, 1989), 68.
38. Malabou, *The New Wounded*, 140.
39. Ibid., 44.
40. Malabou, *The Future of Hegel*, 153.
41. Malabou, *Plasticity at the Dusk of Writing*, 33.

42. Malabou, *Ontology of the Accident*, 37–38.
43. Malabou, *The New Wounded*, 152.
44. Malabou, *The Future of Hegel*, 163.
45. Malabou, *Ontology of the Accident*, 54
46. Ibid., 61.
47. Ibid., 75.
48. Malabou, *Plasticity at the Dusk of Writing*, 9.
49. Malabou, *Ontology of the Accident*, 89.

BIBLIOGRAPHY

Butler, Judith. *The Psychic Life of Power: Theories in Subjection*. Palo Alto, CA: Stanford University Press, 1997.

Granel, Gérard. "Le concept de forme dans 'Das Kapital.'" In *Granel, l'éclat, le combat, l'ouvert*, edited by Jean-Luc Nancy and Élisabeth Rigal, 21–35. Paris: Éditions Belin, 2001.

Malabou, Catherine. *The Future of Hegel: Plasticity, Temporality and Dialectic*. Translated by Lisbeth During. London: Routledge, 2004.

———. *What Should We Do with Our Brain?* Translated by Sebastian Rand. New York: Fordham University Press, 2008.

———. *Plasticity at the Dusk of Writing: Dialectic, Destruction, Deconstruction*. Translated by Carolyn Shread. New York: Columbia University Press, 2010.

———. "Unbind Me." In *A Companion to Hegel*, edited by Michael Baur and Stephen Houlgate, 612–24. Oxford: Wiley-Blackwell, 2011.

———. *Ontology of the Accident: An Essay on Destructive Plasticity*. Translated by Carolyn Shread. Cambridge: Polity Books, 2012.

———. *The New Wounded: From Neurosis to Brain Damage*. Translated by Steven Miller. New York: Fordham University Press, 2012.

———. "Go Wonder: Subjectivity and Affects in Neurobiological Times." *Self and Emotional Life: Philosophy, Psychoanalysis, and Neuroscience*, 1–72. New York: Columbia University Press, 2013.

———. "Will Sovereignty Ever Be Deconstructed?" In *Plastic Materialities: Politics, Legality, and Metamorphosis in the Work of Catherine Malabou*, edited by Brenna Bhandar and Jonathan Goldberg-Hiller, 35–46. Durham, NC: Duke University Press, 2015.

———. "Whither Materialism? Althusser/Darwin." In *Plastic Materialities: Politics, Legality, and Metamorphosis in the Work of Catherine Malabou*, edited by Brenna Bhandar and Jonathan Goldberg-Hiller, 47–60. Durham, NC: Duke University Press, 2015.

———. *Before Tomorrow: Epigenesis and Rationality*. Translated by Carolyn Shread. Cambridge: Polity Press, 2016.

Serres, Michel. *Detachment*. Athens: Ohio University Press, 1989.

Chapter 6

Changing (Reading) Habits—Rereading Hegel Speculatively with Malabou

Isabell Dahms

What is at stake in Hegelian speculation today? This chapter will sketch an answer by analyzing Hegel's concept and its resonances from three different angles. The first part of this chapter is concerned with the notion of speculation in Hegel's thought. While the concept of "speculation" appears consistently throughout all of Hegel's works, this chapter will focus on the unfolding of speculation in the "Anthropology" section on habit, in the *Encyclopedia of the Philosophical Sciences*. This dual focus on speculation and habit will allow us to read Hegel and Malabou together. Malabou's account of habit, it will be argued in the second part of this chapter, both explains and complicates the Hegelian concept of speculation. Her discussion on habit in *The Future of Hegel* allows Malabou to put forward her own understanding of speculation, namely, a speculative strategy of reading and writing philosophical texts that is announced more specifically as a feminist strategy in *Changing Difference*. In view of this feminist turn, we might wonder to what extent Malabou's concept of speculation is still Hegelian. We might also wonder how her own strategy of engaging Hegel differs from other approaches, such as Derridean deconstruction and Irigaray's philosophical mime, from which it tries to distance itself. Can her work announce the potential for a feminist speculation, a new strategy that will offer another way to read and navigate the philosophical canon? These questions constitute the focus of the final section. This chapter's intention is twofold: on the one hand, it aims to develop, with reference to Malabou's work, new readings of Hegel's concept of speculation that reveal the ambiguity that this term has in Hegel's own writings. On the other, in revisiting this concept, it will indicate speculation's potential as a strategy of reading the canon differently and thus of changing reading habits.

PART I

Speculation, in Hegel's writings, is opposed to various other forms of knowing, such as ordinary thought, empirical science, judgment, and perception. Because it is the only form of knowing that can, according to Hegel, hold on to contradiction and, in this way, remedy thought's own defects, speculation, is to become the thinking proper to philosophy.[1] It is this problematic, the role of contradiction not only for thought but also for being, that is discussed again and again in Hegel's writings. And, yet, despite this emphasis on speculation, as that which addresses the question of contradiction, a straightforward definition of this concept is never properly outlined. Instead, speculation is continuously set up against something other and it is by means of this opposition that Hegelian speculation takes shape. Although presented by way of contrast, speculation is not entirely distinct from its counterparts. It is not simply the opposite of reflection, ordinary thought, or the empirical sciences. Rather, speculation is to be understood as a variant of its multiple others, as for instance "a species of reflection."[2] Speculation and its opposites share, according to Hegel, "a certain community of nature."[3] Yet speculation possesses, in addition to the forms shared in common, some forms or features of its own.

There are, following Hegel, always several sides to every entity and every entity, or form of being, coincides with a particular and corresponding form of knowing. For Hegel, this means more concretely that every entity has an abstract side, which corresponds with the understanding, a dialectical side, also termed negative reason and, finally, a speculative dimension or positive reason. These sides do not constitute entirely separate parts of thought and being but are, following Hegel, stages or moments of every logical and existing entity, of every notion and every truth.[4] This multidimensionality of both being and thought explains the simultaneous closeness and tension that we encounter between speculation and the various forms of knowing it is opposed to. Having defined speculation at the beginning of the *Science of Logic* as distinct, though not entirely different, from other forms of knowing, Hegel then goes on to insist on this distinctiveness. The divide between speculation and its other gradually becomes more and more definitive until two quite distinct domains are opened up: the speculative domain in which the *Aufheben* (act of sublation) operates by way of working through contradictions, and a mechanical or formal domain, which understands contradictions to be external and thus calculates only moments of force.[5] In this second domain, the domain of empirical science, ordinary thought, and the understanding, thought is restricted. This is because the formal or abstract nature of mechanical relations does not, according to Hegel, lead to self-reflection and so in this domain the limits of thinking are not overcome. In other words, no speculative dimension is produced.

Although he never explicitly contrasts mechanism and speculation, it is striking that "the mechanic" or "mechanical" appears continuously throughout Hegel's works as a description of that which in philosophy, and in descriptions of natural life and social relations falls short of speculative thought. Hegel refers, to name just a few examples, to mechanical relations,[6] mechanical behavior,[7] and, on the social and political level, to the machine-state, which is not yet a proper political organon and thus only constitutes an atomic, lifeless multiplicity.[8] Since the characterization of the speculative is negative—we only learn about speculation through explanations of its counterpart, through what it is not—these references to mechanism and the mechanical are not at random. In the examples mentioned, Hegel associates mechanism with immediacy, chance, and contingency.[9] He employs this term to refer to the exteriority of certain relations that appear on the surface only but do not constitute the inner life of things.[10] Outer mechanical difference is opposed to internal organic relations. As counterpart of the organic, the mechanical signifies for Hegel merely an aggregation or accumulation, rather than an animate living system.[11] It is the accidental in life, that is, in other words, that which is not yet conceptual or ideal. As such, the mechanical is a state, which, in Hegel's words, manifests an inner meaninglessness.[12]

The notion of the mechanical occupies a peculiar place within Hegel's work because it is both a specific moment in the *Logic* and *Encyclopedia* but is also a term that is employed more broadly as a trope of externality. The mechanical or the exteriority of mechanical relations is set up by Hegel to comprise all that has been demonstrated to be insufficient in the course of the development of the science of philosophy—all the determinations of thought and being that are calculable but are still inadequate. Yet, this characterization of mechanism, as strategic trope, is however not always successful. The ambiguous relation between mechanism and speculation becomes apparent at various places in the *Encyclopedia*, one of which is Hegel's short account of habit in the "Anthropology" section. Here mechanical processes will turn out to be far more complicated than their immediate characterization. It is this section that is also one of the focus points of Malabou's analysis in *The Future of Hegel*. For our reading of Hegel, it is, as Malabou indicates here, critical to make habit a turning point.[13] This might seem surprising at first, given that Hegel's account of habit is rather brief. Introducing this term in the third part of his *Encyclopedia of the Philosophical Sciences*, in the *Philosophy of Mind*, habit appears to be just another fleeting moment. And in a sense, this is true. Hegel, in this section, directs us quickly from one point to another. He however also remarks that almost all scientific works on the mind and soul fail to give us an accurate account of habit.[14] Habit in these works is regarded as not worthy of discussion. Its meaning, Hegel writes, is not understood. What, then, is the significance of habit? And what is its implication for speculation? In other words, how are these two operations, habit and speculation, connected?

PART II

Hegel develops his account of habit in the Anthropology of the *Encyclopedia*, which, in its "strange beauty"[15] brings to light the formation of spirit through man. The Anthropology as a whole and habit in particular manifest the transition from the *Philosophy of Nature* to the *Philosophy of Spirit*. The section on habit in the Anthropology does not unfold straightforwardly as a discourse on habit but begins rather by discussing a particular feeling, the sense of self, which serves as prologue to habit and to the relation between mechanism and speculation that habit displays. The sense of self, as outlined by Hegel, differs from other ordinary feelings or sensations, such as desires, drives, and passions.[16] However, not unlike these other feelings, the sense of self is dependent on a relationship between the soul and body. Not denying the importance of the material or biological body, Hegel nevertheless suggests that a certain break with the mere facticity of the body has to be initiated in order for a sense of selfhood to emerge. The sense of self accordingly is neither somatic nor does it simply refer to the soul. Rather, by signifying a break with the immediate facticity of the body, the sense of self signifies a process of self-differentiation. The soul's reflection on its own corporeality is, however, not yet the reflection of an I or consciousness, though the basis for consciousness has now been established.[17] It is at this point that Hegel mentions habit for the first time.

Habit, according to Hegel, is the dialectical interplay of soul and body and as such is to be understood as the "mechanism of self-feeling."[18] Habit allows for is abstraction. The soul is free by way of abstracting from particular bodily feelings.[19] By way of abstraction, its own body becomes for the soul an outside from which it can reflect upon itself. This is why Hegel suggests that in the mode of habit the soul has the body as its property.[20] The soul itself however is not a property but rather the activity of making one's body one's own. Habit is not only an instrument of abstraction but is at the same time a mechanism of repetition. If these two movements, abstraction and repetition, operate simultaneously, then, Hegel suggests, the exteriority of the body, but also that of the soul can be suspended. Body can be translated into soul. Doing so, the soul comes into existence and is, at that very moment, translated back into the body. A circular reciprocity or causality is established which puts the auto-organization of the body-soul complex into place. It is, Hegel argues, this mechanism consisting of abstraction, repetition, and translation, an unconscious and automated mechanism of the body and soul, that gives rise to an impression of selfhood, that is, to consciousness. A being, Malabou notes, is now given the impression of its existence as something continuous.[21] Habit installs a new immediacy, or second nature that is no longer merely bodily; a second nature that keeps changing so that the self is formed only by being transformed. Yet, while habit awakens consciousness, it also puts it to sleep.[22] As continuous repetition, habit causes detachment in a double sense, a detachment that constitutes the distance necessary for the vitality and persistence of subjectivity but also a detachment that leads to the dulling of life.[23] It follows that conscious and unconscious processes are initiated by the same mechanism, which maintains the self by means of change and repetition of change.

Hegel's definition of speculation in terms of a series of oppositions that culminate in the juxtaposition between the mechanical and the speculative is being complicated in the section on habit, as just outlined. If speculation is ultimately opposed to the mechanical, then how can habit, defined by Hegel as "the mechanism of self-feeling,"[24] give rise to an impression of selfhood, to consciousness and, ultimately, also to spirit? In other words, if habit belongs to the mechanical domain, then how can it be, as Malabou demonstrates in *The Future of Hegel*, a turning point? Substituting by means of repetition one immediacy for another and thus positing a repeated immediacy or second nature, it seems that habit is producing Hegel's speculative dimension. If this is the case, then what is at stake in this distinction between speculation and mechanism? Does it really hold?

Mechanism proves to be an ambivalent element within Hegel's system. While it is argued in the chapter on mechanism in the *Science of Logic* that the properly speculative is still lacking in mechanism,[25] Hegel however acknowledges elsewhere, as for instance in the Anthropology section of *The Encyclopedia*, that there are mechanisms, such as habit and memory,[26] that fulfill vital functions. In the *Science of Logic* mechanism also turns out to be more complicated than its immediate characterization. Hegel will here even go so far as to acknowledge that mechanical operations involve spontaneous activity, impulse, and some form of consciousness.[27] If this is the case, then what is lacking in the mechanical as such? What divides the mechanical from a more advanced speculative relation? According to Hegel, a certain freedom remains lacking in the mechanical mode. Relations that should be internal to objects remain here to a certain extent "exterior." The relation to self, for instance, proves to be a "merely mirrored relation"[28] that is determined by another. It follows that the relations that characterize and determine an object have not yet been truly internalized and are thus not entirely its own.[29]

Commentators, such as Luce Irigaray and Jacques Derrida, have challenged Hegel on this point. In *Speculum of the Other Woman*, Irigaray calls our attention to the etymological origin of the concept of speculation. She points out that the two terms, speculation, which initially meant to contemplate and to obverse, and the mirror, reflector, or speculum have the same linguistic origin. Both terms, as Irigaray demonstrates, derive from the Latin *specere*, meaning to look at or to view. We might therefore wonder with Irigaray, whether it is not at this point, in the mirror stage of mechanism, that we find ourselves unexpectedly back in speculation. If the mechanical is a mirrored relation, then are we here not already in some speculative domain? Derrida raises a similar objection to Hegel's presentation of the speculative. According to him, the exposition of mechanism and speculation as opposites leads to two main difficulties. If the divide between the speculative and its other(s) is maintained, then, Derrida argues in *Glas* and *Writing and Difference*, the speculative domain will be one of exclusion, a closed and restricted economy or discourse that rejects its outside. However, if the difference is overcome dialectically, which is to say, sublated from the standpoint of the absolute or of speculative reason, then another problematic emerges. What if some mechanisms turn out to be too complex to be sublated? If certain mechanical operations like habit initiate at the same time conscious and unconscious processes and thus give rise, in their complexity, to an impression of selfhood, then we have to examine whether the mechanical per se can simply be interiorized by the speculative as one single moment. We would have to reconsider whether sublation is the right term in order to capture the relationship between the mechanical and speculation. Sublation, which means at the same time to suppress and to preserve, arguably differs from the exchange or translation that is the work of habit, and from the reduplication of the mirror, both of which are mechanical operations, according to Hegel.

The second difficulty, as addressed by Derrida, leads us back to the first: speculation excludes its outside. Because Hegel indicates that the mechanical can simply be assimilated by spirit as its negative, Derrida concludes that Hegelian speculation restricts us to a world of prohibition.[30] The mechanic and everything mechanical—habit, memory, the abstract and exterior relations and modes of being—seem to become irrelevant once the speculative has come into view. What is excluded in this restricted economy of Hegelian speculation is another type of relation other than sublation that can link the elements of the system. Hegel restricts us to one particular mode of exchange, sublation, whereas all other modes, though preserved, are, to borrow Irigaray's words, hidden in the dark cave where they serve as secret foundations to speculation.[31] Hegel, following Derrida, resolves all ambiguity that can be found within in his system by setting up a hierarchical order in which speculation is opposed to one-sidedness, while everything associated with mechanism is, by contrast, only a partial aspect of the true. Instead of retaining different forms of relation, such as sublation, translation and mirroring, as equally necessary, Hegel seems to settle for one, sublation. This allows him to quickly pass from one element of his system to another, for example, from habit (mechanical) to consciousness (speculative) or from memory (mechanical) to thought (speculative), without exploring the possibility of a negotiating relationship between the two in sufficient depth. While all mechanisms pass over into their other, Hegel claims that the speculative, by contrast, does not disappear. Speculation, then, is the only term that carries itself into effect.

If we return to Malabou's reading of habit in *The Future of Hegel*, a different interpretation becomes possible. Malabou concludes that speculation does not always operate in the mode of sublation. Habit is a turning point because, as Malabou proposes, the translation of body into soul, brought about by the mechanism of self-feeling, is also the form or structure of speculative meaning.[32] What Malabou seems to suggest is that habit is the ontological expression of a logical structure, a structure termed speculation by Hegel. The logical element, Malabou argues, is the primal environment of thought and thus of philosophy; but thought, like life, possesses the power of determining and embodying itself.[33] This is what happens in the Anthropology where the logical process is translated into a sensuous form.[34] Without this incorporation, Malabou writes, the speculative would be ineffective and logical life only an abstraction.[35] This is why the role of habit and of mechanism is crucial within the Hegelian system. Moreover, since habit occurs not as a sublation, but as a translation and reduplication, it would seem that Hegel's speculative resolution does not have to, each time, evolve in the same manner. Rather than just one mode of exchange, others also come into view. Passing from mechanism to speculation, the elements of the system get translated, mirrored, reduplicated, and sublated.

Is such a reading of Hegel plausible? Replacing sublation as the central element of speculation, what we are left with is the dialectical interplay of mechanism and speculation, a union without fusion of two equally important operations, as predicted by Malabou in *The Future of Hegel*.[36] Does this mean that the dialectic is what wins out in the end, over and beyond mechanism and speculation? Certainly not for Hegel. It almost seems as if, according to Hegel, dialectics on its own accord is too mechanical—a natural immediacy, or one more accidental habit that has become an essence of being. With speculation another dimension is added, less predictable in its procedure and in this sense less mechanical, though utterly dependent on mechanisms to sustain its operation. With speculation, a certain fragility enters the system.[37]

PART III

What then is at stake in Hegelian speculation today? At stake for Malabou is to complicate these categories—the speculative and the mechanical—and the modes of relation between them. It is arguably in this movement from the one to the other that we come to understand the dynamics of the Hegelian text. It is also here that Malabou's own intervention and her proposition of a speculative strategy of reading and writing philosophy is to be situated. We cannot, as Malabou notes, rediscover Hegel intact, untouched by deconstruction, by critical theory, feminist or postcolonial thought. However, by recovering the different modes of relation between Hegel's key terms and categories, a speculative reading can take shape and some form of discovery becomes possible that goes beyond a mere understanding of the text. In this sense, reading Hegel consists not only in following carefully the logic of argument; it also consists in grasping the accidental.[38] Hegel's philosophy, Malabou writes, "announces that the future, from now on, depends on the way the shapes and figures already present can be put back into play, on the way the extraordinary and unexpected can only arise out of the prose of the well-known and familiar."[39] Emphasizing both the transformability of the text itself and the transformative power of the reader, Malabou argues that thought, due to its speculative nature, renounces the rigidity of its original position.[40] Not only is reinvention possible, it ought to happen and it, moreover, does not mean to repeat, to mime, or to reproduce.[41] Malabou therefore suggests relaunching, beyond deconstruction, authors like Hegel and philosophical categories like speculation that were thought complete, over or worn out by tradition.[42] In *Changing Difference*, where this strategy becomes a feminist one, Malabou emphasizes that her speculative reinvention is not Irigaray's strategy of mimicry and is not Derridean deconstruction either. If this is the case, then what kind of recovery is to be found in Malabou's Hegelian intervention? What is being invented in this speculative discovery?

What is to be found in the interrelations between the key terms of a text is, Malabou suggests in *Changing Difference*, a structural potential for change. What gets reinvented are the sites where the philosophical text forms itself, that is the sites where its key elements are renegotiated. In reinventing these, there is arguably a great deal at stake. It is at this point that the possibility of a feminist reading comes into view. Can Malabou's speculative strategy offer new ways to read and navigate the philosophical canon? If concepts can be relaunched, speculatively, in thinkers like Hegel, then what Malabou asks us to do is to change our habit of reading them. While habit, as outlined, is a mechanical operation, there is also a speculative aspect to it when habit overcomes its own self to create new habits and to form a self only by way of transforming it. As such, habit creates new ways of being at home in the world. This is a feminist concern. This concern demands the questions: How to make oneself at home in the discipline of philosophy, how to relaunch its concepts and texts, to turn them against themselves so that they can address new questions?

Complicating one of Hegel's key terms by reading speculation alongside habit and mechanism, we unsettle his system. But we will also find that the system never truly closes itself, as Malabou and other readers of Hegel, including Nancy, Adorno and Derrida, have shown. It is for this reason that we can reinvent, that is reread traditional philosophical concepts. We might however wonder how far we can go with such a speculative rereading. If theories that are critical of the canon, such as feminist philosophies, critical race theory, deconstruction or the work of the Frankfurt School have shaped our philosophical understanding, then what implications does this have for our reading of Hegel? Do we have to mercilessly challenge canonical texts like Hegel's and their entrenched norms, which are often patriarchal and racist? The question is how to negotiate the need for critique with the possibility of inventing anew. Several strategies to navigate the Hegelian text are already available to us. We might come to decide like Carla Lonzi and *Rivolta Femminile* to spit on Hegel because he, among others, has justified by means of metaphysics the reality of oppression of women,[43] of people of color and particularly women of color. By not recognizing ourselves in his system, we might be able to deprive it of the illusion of its universality.[44] To do so, we could also play with *mimesis*, as Irigaray suggests. Mimesis, according to Irigaray, is an attempt by women to locate their place of exploitation by discourse without allowing themselves to be simply reduced to it.[45] This means for women to submit themselves to ideas about themselves, elaborated within a masculine logic in order to make visible from within, by an effect of playful repetition, what was supposed to remain invisible, namely, that this logic is neither neutral nor unchangeable.[46] Another possibility would be to intervene more directly and to strategically deconstruct the absolute in all its forms and expressions, as well as Hegelian speculative syntheses, whenever they occur – a strategy that Derrida performs in *Glas*.

Malabou too reflects, in *The Future of Hegel*, on the question of how to read Hegel today. In *Changing Difference* this question is broadened and at the same time becomes more specific. Malabou now asks in what way and how far philosophy, and not just Hegel, can transform itself under the impact of women's resistance.[47] It is this question that turns into an alternative strategy, one that is neither deconstruction nor mimesis. The reader, however, might wonder what are the problems with miming, deconstructing and spitting on Hegel? In response to the latter approach, some feminist philosophers, for instance Christine Delphy, have argued that feminist thought is not something entirely different from other discourses and that thus some productive form of encounter will be needed. Feminism, arguably, does not involve a total break with other discourses, for that would entail cutting back the content of feminist discourses considerably, possibly leaving it unintelligible.[48] Since "no movement of ideas is a pure opposition, a pure denial,"[49] deconstruction and mimesis might be more productive in the long run. However, while Derrida's deconstruction has offered theoretical tools to dismantle the "universal" subject of philosophical propositions, a subject that by default has been white and male, he has also with that very same gesture, Irigaray and Malabou argue, undermined more affirmative potentials for resistance.[50] Thus, while Derrida has undoubtedly broadened our potentials to think through philosophical concepts critically, uncovering their underlying assumptions, this strategy in itself does not yet constitute a political act. It might even pose "a total threat to feminism."[51] By problematizing binary oppositions and forms of identity, Derrida, Malabou claims, leaves women without any essence.[52] Irigaray voices a similar concern. The deconstruction of identities, if expressed simply as a logical-theoretical demand, continues, according to her, to leave women in a state of fragmentation and dissemination, which reproduces and thus perpetuates patriarchal violence.[53]

Malabou therefore asks whether women must, despite all their efforts, always remain running behind, resigning themselves, pretending, inhibiting their energy, installing themselves in mimicry, forever enveloping themselves in clothes that are not made for them.[54] This question is addressed to Derrida but even more so to Irigaray. Not an outspoken critic of the tactics of mimicry, Malabou yet suggests that such a strategy might become tiring after a while and ultimately unsatisfying. Irigaray's philosophical mime, Malabou observes, runs the danger of getting stuck in repetition.[55] Once the place of exclusion has been identified, it might no longer be a question of contesting a preestablished value system through mimicry, but of finding ways not to repeat this system.[56] Mimicry, Malabou observes would after all change nothing about the fact that all the "philosophical" questions and concepts are always given to women and are thus borrowed topics.[57]

While the strategies of total critique, of deconstruction and mimesis are emancipatory in their own regards, another way might be possible. In Malabou's case, this other strategy consists in attaching the fate of philosophy to the fate of the concept of "woman." The plasticity of philosophy, she writes, and the plasticity of "woman" must be put to the test together.[58] Malabou is not prepared to give up on the philosophical canon and its concepts. She will remain a Hegelian after all. For her, to transgress the limits of philosophy and its violence does not necessarily imply giving up on concepts—the terrain of philosophy—which do not carry the burden of phallogocentrism.[59] For the ideology at work in philosophical concepts, Malabou writes, does not come from the concepts themselves, but from their situation and employment in philosophical speech or writing.[60] Here, one could argue, Malabou is close to Irigaray again. Her tactic of a speculative rereading and reinvention seems to rely on the insights of the mimetic philosopher. According to both Irigaray and Malabou, by remaining within the ambivalence of the canon, miming and inventing, it becomes possible to relaunch what was thought complete. Given her indebtedness to Irigaray and Irigaray's own influence and involvement in the Italian feminist movement, a return at this point to Italian feminist practices of reading and writing might be productive. Not only Carla Lonzi but also Luisa Muraro, Adriana Cavarero, and the works of the group Diotima and The Milan Women's Bookshop Collective could be of interest to Malabou's Hegelian attempt at changing reading habits.

Reading Hegel speculatively with Malabou, we are accordingly less interested in Derrida's opening question in *Glas*—what remains for us of Hegel today[61]—and more invested in the question of reinvention. What gets reinvented and by whom? This is a philosophical question but a political one as well. It is an invitation to read philosophy and, in doing so, to change the habits of this discipline. This, a tactic among others, is what Malabou's feminist speculation might look like. While the plasticity of the concept of "woman" has yet to be further specified by Malabou, a first demand has been made. "Woman" is no longer to signify the "nonplace," the ever-deferred ontological promise, the empty essence and outside of philosophy. "Woman" is not to remain a philosophical concept that can give no particular place to no particular woman.

NOTES

1. G. W. F. Hegel, *Enzyklopädie der philosophischen Wissenschaften* (Hamburg: Felix Meiner Verlag, 2015), 49.
2. Ibid., §9.
3. Ibid.
4. Ibid., §79.
5. Francoise Balibar, Phillipe Büttgen, and Barbara Cassin, "Moment, Momentum, Instant," in *Dictionary of Untranslatables*, ed. Barbara Cassin, Emily Apter, Jacques Lezra, and Michael Wood (Princeton, NJ: Princeton University Press, 2014), 686.
6. Hegel, *EL*, §291.
7. Ibid., §295.
8. G. W. F. Hegel, *Jenaer Kritische Schriften* (Hamburg: Felix Meiner Verlag, 2015), 58.
9. Hegel, *EL*, §310.
10. Ibid., §445.
11. Ibid.
12. Ibid., §187.
13. Catherine Malabou, *The Future of Hegel: Plasticity, Temporality and Dialectic,* trans. Lisbeth During (London: Routledge, 2005), 25.
14. Hegel, *EL*, §410.
15. Malabou, *The Future of Hegel*, 23.

16. Hegel, *EL*, §409.
17. Ibid.
18. Malabou, *The Future of Hegel*, 25
19. Hegel, *EL*, §410.
20. Ibid.
21. Ibid., 64.
22. Malabou, *The Future of Hegel*, 76.
23. Ibid.
24. Hegel, *EL*, §410.
25. G. W. F. Hegel, *The Science of Logic*, trans. George di Giovanni (Cambridge: Cambridge University Press, 2010), 631.
26. Memory is defined as the mechanism of intelligence.
27. Hegel, *SL*, 631.
28. Ibid.
29. Hegel, *EL*, §12.
30. Jacques Derrida, *Writing and Difference* (London: Routledge, 2001), 348.
31. Luce Irigaray, *Speculum of the Other Woman* (Ithaca, NY: Cornell University Press, 1985), 144.
32. Malabou, *The Future of Hegel*, 69.
33. Ibid., 139.
34. Ibid., 74.
35. Ibid., 140.
36. Ibid., 69.
37. For a more detailed account of speculation as fragility, as what opens philosophy and thought to the experience of aporia and brokenness, see also Gillian Rose's work, for instance *Hegel Contra Sociology*, and Howard Caygill's interpretation of Rose's retrieval of speculative philosophy in "The Broken Hegel: Gillian Rose's Retrieval of Speculative Philosophy."
38. Jean-Paul Martinon, *On Futurity: Malabou, Nancy and Derrida* (Basingstoke, UK: Palgrave Macmillan, 2007), 37f.
39. Malabou, *FH*, 190.
40. Ibid., 142.
41. Ibid., 73.
42. Catherine Malabou, *Changing Difference: The Feminine and the Question of Philosophy*, trans. Carolyn Shread (Cambridge: Polity, 2011), 72f.
43. Rivolta Femminile *Manifesto* (1970): http://www.columbia.edu/itc/architecture/ockman/pdfs/feminism/manifesto.pdf.
44. This is the announced strategy and aim of both the Rivolta Femminile *Manifesto* and Irigaray's *Speculum of the Other Woman*. The chapter in *Speculum* that focuses directly on Hegel is entitled "The Eternal Irony of the Community."
45. Luce Irigaray, *This Sex Which Is Not One* (Ithaca, NY: Cornell University Press, 1985), 76.
46. Ibid.

47. Malabou, *Changing Difference*, 110.

48. Christine Delphy, "Changing Women in a Changing Europe: Is 'Difference' the Future for Feminism?" in *Women's Studies International Forum* 17, no. 2–3 (1994): 187.

49. Ibid.

50. See Derrida's "Women in the Beehive" and feminist critiques of Derrida's appropriation of the feminine, for instance, by Irigaray in *Marine Lover*, by Malabou in *Changing Difference* but also, though hers is a much more generous negotiation with Derrida, by Gayatri Chakravorty Spivak, for instance, in "Displacement and the Discourse of Woman" in *Feminist Interpretations of Jacques Derrida* and "Feminism and Deconstruction, Again: Negotiations" in *Outside in the Teaching Machine*.

51. Malabou, *Changing Difference*, 107.

52. Ibid., 109.

53. Margaret Whitford, *Luce Irigaray: Philosophy in the Feminine* (London: Routledge, 1991), 123.

54. Malabou, *Changing Difference*, 110.

55. Ibid., 129.

56. Ibid., 133.

57. Ibid., 102.

58. Ibid., 110.

59. Ibid., 110f.

60. Ibid.

61. See Derrida's *Glas* (Lincoln: University of Nebraska Press, 1986), 1.

BIBLIOGRAPHY

Balibar, Francoise, Philippe Büttgen, and Barbara Cassin. "Moment, Momentum, Instant." In *Dictionary of Untranslatables*, edited by Barbara Cassin, Emily Apter, Jacques Lezra, and Michael Wood, 683–89. Princeton, NJ: Princeton University Press, 2014.

Delphy, Christine. "Changing Women in a Changing Europe: Is 'Difference' the Future for Feminism?" In *Women's Studies International Forum* 17, no. 2–3 (1994): 187–201.

Derrida, Jacques. *Glas*. Lincoln: University of Nebraska Press, 1986.

———. *Writing and Difference*. London: Routledge, 2001.

Hegel, G. W. F. *Enzyklopädie der philosophischen Wissenschaften*. Hamburg: Felix Meiner Verlag, 2015.

———. *Grundlinien der Philosophie des Rechts*. Hamburg: Felix Meiner Verlag, 2015.

———. *Jenaer Kritische Schriften*. Hamburg: Felix Meiner Verlag, 2015.

———. *The Science of Logic*. Translated by George de Giovanni. Cambridge: Cambridge University Press, 2010.

Irigaray, Luce. *Speculum of the Other Woman*. Ithaca, NY: Cornell University Press, 1985.

———. *This Sex Which Is Not One*. Ithaca, NY: Cornell University Press, 1985.

Malabou, Catherine. *Changing Difference: The Feminine and the Question of Philosophy*. Translated by Carolyn Shread. Cambridge: Polity, 2011.

———. *The Future of Hegel: Plasticity, Temporality and Dialectic*. Translated by Lisbeth During. London: Routledge, 2005.

Martinon, Jean-Paul. *On Futurity: Malabou, Nancy and Derrida*. Basingstoke, UK: Palgrave Macmillan, 2007.

Rivolta Femminile. "Manifesto." Last Modified March 29, 2002. Accessed September 25, 2017. http://www.columbia.edu/itc/architecture/ockman/pdfs/feminism/manifesto.pdf.

Whitford, Margaret. *Luce Irigaray: Philosophy in the Feminine*. London: Routledge, 1991.

Chapter 7

Habitués[1]

Thomas Wormald

This chapter will pursue an interpretation of habit different than the contributions of Sandrine Hansen and Isabell Dahms, and different than the one originally elaborated by Catherine Malabou. While most commentaries on or readings of Malabou and habit rightly focus on her original reading of the central role of habit in Hegel's "Anthropology" in *The Future of Hegel Plasticity, Temporality and Dialectic*, other sites remain in the Hegelian corpus—principally the *Philosophy of Right*—where habit assumes different, potentially consequential inflections. I wish to tease out or pursue these interpretive possibilities in order to clarify what is ultimately at stake in the concept of 'habit' when we sometimes benignly or unreflectively refer to the various habits that structure our lives: socio-political, philosophical, onto-epistemic and otherwise. Malabou's reading of habit in *The Future of Hegel* is principally concerned with the moment of subjective spirit, the becoming of the body and "I," which is to say that it is principally a reading of an individual process or phenomenon. Using Malabou's interpretation of habit as a departure point, I want to extend or transpose her insights and elaborate their consequences for a reading of habit as it appears in the *Philosophy of Right*, in the context of objective spirit or sociopolitical life, our relation to and with others, the constitution of a "habitus" and what this entails, and, secondly, the philosophic habits that sustain or structure this life: our ontoepistemic habits or our ways of thinking and doing. I argue that habit assumes deeper significance if we shift vantages to consider its implications from these perspectives. One of the hallmarks of Malabou's reading is the insistent emphasis and exploration of the speculative meaning of habit, its simultaneous operation as a mechanism of both life and death, of possibility and impossibility, of change and stasis—yet, these reflections remain, within the remit of her explication, at the individual, ontological level. At this stage or level, habit essentially concerns our acclimation to our own body and the successive shaping of ourselves through the internalizing adoption and adaptation of various externalities (skills, capacities, etc.) via the naturalization of the "unnatural," the creation of a second nature. From this, Malabou importantly obtains a materialist account of subjectivity and an exemplification of plasticity: how the subject fashions and refashions itself through a processual, temporally ecstatic reception and (self) donation of form. Yet, if we shift this analysis to a different register, to that of the construction and maintenance of a "habitus," that is, of society, of our living-with, and to the habits of thinking that subtend, structure, and sustain this world, we encounter a different set of concerns than a dialectic of alienation and incorporation at the level of the self, instead confronting a problematic of the structuring relation of forms of life and the world and the constitution of "home" in which we can be at one with. In this context, Malabou's analysis of habit as a plastic force of both life and death, of possibility and impossibility, of being or not being at home, become questions of: what forms of life live and die, what forms of life are rendered as possible and impossible, what forms of life are at home in the present shape of the world and which are effectively rendered the noninhabitants, those not-at-home in the world?

First, I will give an overview of Hegel's account of habit in the moment of subjective spirit at the level of the self, subject or individual, stressing Hegel's description that it is a way to be-at-home-with-oneself, then highlight the key points of emphasis in Malabou's reading of habit. Specifically, I will isolate the unique accents in Malabou's reading of habit: habit as a kind of property or ownership, habit as a kind of productive amnesia, and the fascinating final prescriptions regarding habit at the end of *The Future of Hegel*: the role of philosophy with respect to our habits of thought and, second, the slightly cryptic observation that the future depends on the "habitués" looking at their own habits. Using these features of Malabou's analysis, I will then use them to extrapolate and speculate on the consequences of what they mean in the context of habit at the level of objective spirit, the constitution of our shared sociopolitical life and the philosophical or ontoepistemic practices that sustain, structure, or in-form this world. If habit is the making of one's own, making a home for oneself, what does this mean in the context of sculpting a world together, making a home-with-others? Focusing on this small phrase closing her *The Future of Hegel*, I will suggest that there is a nascent analytic of habitués and nonhabitués that remains to be elaborated out of Malabou's thought.

HEGEL ON HABIT IN SUBJECTIVE SPIRIT

The trajectory of the "Anthropology" in the subjective spirit division of the *Philosophy of Mind* traces out the development of the self through its gradual habituation to its own body to the point at which a conscious "I" or subject emerges from its natural or substantial ground. This trajectory follows the typical path of Hegel's triadic dialectical thinking, moving from the universal, to the particular, to the singular. One begins with The Natural Soul, which comprises a "universal substance" that each individual inherits and which becomes the subsequent material basis of all individual sculpting and formation.[2] By Natural Soul, Hegel means the various accidental features of one's life that one is born into, such as geography, nationality, and ethnicity. The next moment is what Hegel calls "The Feeling Soul" which itself moves through the triadic moments of universal—particular—singular: the feeling soul in its immediacy (universal), self-feeling (particular) and, finally, habit (singular). The feeling soul in its immediacy can be described as the subject being "in itself" and not yet "for itself," still in "darkness"[3] as an "infinite treasury of sensations";[4] "it is . . . immediate, not yet as its self, not a true subject reflected into itself."[5] This is the human being in its sheer corporeity where the feelings and sensations are its mode of being—there is, as of yet, no self-differentiation. The next moment, what Hegel calls self-feeling, constitutes the point at which the self begins to distinguish itself from its feelings transitioning from the *universality* of feeling in its immediacy to the *particular* moment of self-feeling where the individual inchoately begins to feel that these feelings are *its* own: it is "essentially the tendency to distinguish itself in itself, and to wake up to the *judgment in itself*, in virtue of which it has *particular* feelings and stands as a subject in respect of these aspects of itself."[6] In this moment of development, the individual essentially experiences a kind of diremption in where the body and its attendant passions, ideas, inclinations, and capacities appear as externalities instead of aspects of its being; here, the subject, at the level of the particularity of self-feeling, is *alienated* from itself.

The next stage, that of habit, is the corrective to this moment, being the mechanism through which the subject, through a habitual process of practice and repetition, comes to integrate these accidents, contingencies, or particularities as features of its own identity. Through the work of habit, the self internalizes these different externalities and comes to be individuated and at home in them. Habit first appears as the phenomenon through which the body forms itself by "reducing the particulars of feelings (and of consciousness) to a mere feature of its being."[7] This is the process whereby the natural soul incorporates into itself the initially alien particularities of sensation so it "has them and moves in them, without feeling or consciousness of the fact."[8] The emergence of the subject, for Hegel, is subtended by this plastic movement of the body autoorganizing itself through the contraction of habits, as the body's corporeality is molded without rational intervention or consciousness. This operation of habit is effectively the sculptor of the marble of plasticity: matter actively forms itself and the self passively acquires form, unconsciously, in the development of its body and bodily capacities. This originary activity and passivity of matter enables the self to take form and exist, as in appropriating and internalizing these externalities of self-feeling, "the soul is freed from them, so far as it is not interested in or occupied with them."[9] Becoming habituated to—that is, at-home-in—the clamorings of its desires and needs and developing capacities and skills, the individual becomes "open to be otherwise occupied and engaged."[10] This phase of habit, through which the body habituates itself into a functioning homeostasis, creates the very possibility of the emergence of a conscious, reflective self and concerted, creative activity. The passive syntheses of habit "form the condition of the soul's liberation, of its attaining objective consciousness,"[11] establishing the stability required for the subject to both participate in future self-directed initiatives and withstand future encroachments from and encounters with its external material environs. When the self finally develops the capacity for reflection and intentional consciousness it enters, decisively, *in media res* of an affective, embodied, biological life already very much long in the making through the unconscious work of habit. Our bodies and their ability to exist and consist in the world are a result of manifold repetitions, practices, and unreflective processes—that is, a host of habits—which we do not choose to participate in but comprise the very conditions of possibility for being an individuated self. Habit comprises or provides us with a kind of unreflective infrastructure by and through which we live. It is the manner by which we overcome the alienation we experience within the very experience of our own bodies and come to be *at home* with ourselves.

MALABOU'S READING OF HABIT IN *THE FUTURE OF HEGEL*

A reading of Hegel's account of habit forms a central part of Malabou's itinerary in *The Future of Hegel*. Hegel's account of habit furnishes Malabou with a surprisingly materialist account of subjectivity, as through habit Hegel elaborates the process whereby the body sculpts itself, emphasizing the prereflective processes prior to consciousness by which the body develops, internalizes, and make's one's own affective, physiological, and cognitive habits that permit the body to attain the functional homeostasis required to be an operative, embodied individual. In this concept Malabou finds "a genuine plasticity,"[12] a speculative concept grounded in an interplay of passivity and activity whereby our "first natures" are refashioned or sculpted into "second natures," accidents into essences, exemplifying our plastic capacity for both transformation and stability; our capacity to both retain changes and experience and welcome the new. Through habit Malabou finds an operation that corroborates that notions such as form and essence are not merely the detritus of metaphysics but require to be contemporarily reelaborated, her investigation into habit confirming the ontological (ex)change between categories like essence and accident, *a posteriori* and *a priori*, contingency and necessity; habit continually disrupts the rigidity of these demarcations, troubling their order of priority.

If these comprise some of the broader stakes of Malabou's interest in habit, there are specific accents that she places on elements of Hegel's account that, we will argue shortly, assume particular consequence if we enlarge the scope or purview of habit to include the other registers in which this concept significantly functions in Hegel: that of thought itself and the relation of self and world. As we have seen, habit is a coming to be-at-home-with-one-self, a way of being achieved through appropriating what initially appears external or alien to us and making it one's own. Malabou draws our attention to this early in *The Future of Hegel*, writing: "in the etymology of the word 'habit' we discover the Latin word *habere*: 'habit' is a way of 'having', and in this sense, a kind of possession, a property."[13] Malabou further glosses this to mean "a particular kind of having which becomes a way of being."[14] Habit then consists in a way of having, of appropriating, internalizing, taking over: it is the making of one's own something that was not previously that becomes a structuring way of orienting oneself in the world, a way of being. In the language of Hegel, it is the capacity for the subject to internalize what it perceives to be an object, or something independent of themselves: to take in that which stands over and against it and integrate it into the being of oneself and come to move freely with it, own it, or be at-home-in and with it.

Second, Malabou highlights the consonance between the work of habit and the work of sculpture, observing that habit's work is "analogous to the moulding gesture of the sculptor."[15] In Malabou's ontology, plasticity is the "marble" of existence[16] through which habit functions as the sculptor, materializing and giving shape to the undifferentiated material at hand. Through the act of chiseling and contouring, the marble of the self is sculpted so as to gain a determinate shape and particular form. The cultivation of habits, it is crucial to recognize, originally constitute acts of will and choice. As Hegel writes, "The form of habit, like any other is open to anything we chance to put into it."[17] While habits subsequently come to be second nature to the subject, they initially require practice, repetition, and thus a certain amount of intention; to create a determinate shape out of a piece of marble requires a purposeful hand. Habit is thus, Malabou writes, "the process whereby the contingent becomes essential,"[18] whereby concerted effort alchemizes into a certain effortlessness. Malabou thus thirdly highlights that habit operates through a process of a kind of productive amnesia: the creation of a form of the self through the work of habit occurs essentially through what she variously describes as "disappearing subject," a "self-absenting," or a "self-forgetting."[19] Something becomes "second nature," becomes part of oneself, when habit effectively takes hold, when it ceases to be independent of the subject and becomes a seamless feature of one's essence or identity. Habit then effaces the history of its own character: it conceals the contingency of its emergence in the guise of "its always been," exhibiting "an innate tendency to conceal its identity as habit."[20] Habit hides itself as habit to (dis)appear to the habituated as something natural: as first and not second nature, not as something that could be otherwise but something that merely "is," as if it were purely "given," emanating from the nature of the world itself. It is precisely through this forgetting by which habit achieves its economy or efficacy. This leads to Malabou's last accent that habit acts at once as a force of both life and death: "Habit murders man. And it does so just as surely as it makes man live."[21] Habit structures and sediments particular ways of acting, being, and thinking to open up other ways to be engaged or occupied, yet this very act of sedimentation risks robbing the subject of the vitality and impetus to activity necessary to realize these pursuits. Habit can capture or overtake the subject, the convenience, facility, or repetitive insistence of the familiar becoming so rote that one cannot summon the capacity to detach from, rearrange, or recompose their habits, rather further sinking to them, devolving into a lifeless, routinized passivity estranged from the possibility of difference or the new. It is in this sense that Malabou states that habit constitutes a "deadening principle," a "murderous power."[22] While habit always harbors the possibility of plasticity because it initially *became*, that is, (ex)changed or transformed from the contingent into the essential, and thus, can change or be put in play again, this mortifying effect always risks putting under duress and imperiling habit's more affirmative and transformative potential. The last consequence to draw here between the relation of habit and death is that because of the nature of habit, its tendency to petrify and recapitulate, because its vitality and reserve of virtuality is always fragile and finite, the development and taking-hold of particular habits means a consequent foreclosure of others: the making possible of some particular habits means that others are made difficult, improbable, even impossible; further still, the life of some habits means, necessarily, the death of others.

With these, we have outlined what we have identified as the salient, distinctive contours of Malabou's elaboration of Hegel on habit, the threads of which we can collect and rearticulate here:

1. Habit is a mark of property or possession; it is the creation of a "home" whereby the self overcomes its diremption or alienation to be at "one" with oneself.
2. Habit is likened to a sculptor: the process by which habits are cultivated is one that is purposeful, taking effort to achieve.

3. Habit consists in a particular productive amnesia: when habit formalizes (takes form), the habituated cannot recollect the contingency of its origin.
4. Habit is a force of life and death: the ossification of habit can outstrip and overtake the vitality and activity it is supposed to make possible; the cultivation of some habits necessitate the foreclosure of others.

We must lastly index Malabou's closing words regarding habit, which amount to a kind of prescription of "what is to be done" with respect to habit. First, with regard to thought, Malabou observes that the capacity and role of philosophy is that of enacting a "necessary anamnesis": "philosophy sparks off a necessary anamnesis that undoes spirit from its alienated relation to the familiar and the well-known, which drives it away from the customs of the *doxa*, thus involving it in the work of recomposing new habits, new hermeneutical habits."[23] Philosophy then, Malabou writes, serves the role of rectifying the amnesia of habit, of fluidifying our fossilized ways of being, thinking, and acting and opening up the space and possibility of creating new, more apposite habits. Continuing, Malabou writes: "Thought's very life depends on its power to awaken that vital energy which tends to 'mortify' itself, to become sedimented into fixed and rigid positions. The outcome that will follow depends on this awakening: *thought has nothing to do but wait for the habitués to look at their habits*."[24] It is this curious, small phrase that I want to ultimately focus on to unpack the consequences of Malabou's reading of habit. A habitué is defined as a regular, a frequent visitor, a familiar patron—a resident, someone welcomed and accustomed. Now we will unfurl the significance of this in the context of philosophical and sociopolitical habits: the construction of a "home" for thought and a home for dwelling; a world in which some are at home, and others are not.

OTHER SCENES OF HABIT: THOUGHT AND WORLD

While Malabou highlights and affirms the role of habit and its relation to plasticity in the ontological constitution of the individual, she gestures to but does not substantially explore the similarly crucial function habit occupies in Hegel's objective spirit in *The Philosophy of Right*, the realm of ethical or shared sociopolitical life. As Hegel writes in the section of subjective spirit which occupies Malabou's sustained attention, "The form of habit applies to all kinds and grades of mental action."[25] That is, the role of habit is not exclusive to the domain of subjective spirit or the constitution of the body and its capacities, but performs a similarly integral role in the more complex, interpenetrating levels of both ontoepistemic practices, that is thought and knowledge production itself, as well as world formation or the relation between self and society. As Hegel writes in the *Philosophy of Right*, "Habituation thus belongs to ethicality [ethical life] just as it belongs to philosophical thinking."[26] If habit means being-at-home-with-oneself in the context of subjective spirit, habit assumes the significance of being-at-home-with-one's-world in the context of objective spirit. Whereas the former concerns the alignment or reconciliation of a split within the subject itself, coming to be at home in one's body, the latter concerns the structuring alignment or reconciliation of the individual and the world, a coming to be at home with the ways of thinking and being that constitute our shared existence. In this dialectic, it is the world and its subtending ontoepistemic infrastructure or what Hegel calls ethical substance that occupies the status of the object of which the subject or the individual stands opposed. Hegel's analysis of habit in the subjective spirit's "Anthropology" is essentially reprised and recapitulated at the level of objective spirit:[27] the individual experiences the world as something initially alien or external to them, and it is the work of habit that effectively heals or reconciles this division. We initially perceive ourselves as particular individuals independent of ethical substance, responding by seeing the "laws and powers" of ethical substance as an "eternal justice, as gods that are in and as themselves, over against which the vain doings of individual humans remain merely an ever-undulating play."[28] At this stage, ethical substance "relates to the subject as something that simply *is*, in the highest sense of self-sufficiency. In this respect, it is an absolute authority and power infinitely more stable than that of nature."[29] Ethical substance, the ways of being and thinking that structure our world, thus confronts the individual as a foreign, alien, oppressive object, an unmalleable stage upon which the individual merely acts: the objective world stands over and against the individual as something which just *is*, that exists independently of the individual's activity and existence.[30] As it was in subjective spirit, it is through becoming habituated, appropriating or internalizing what the individual first experiences as externalities which are in this context ways of thinking and being, norms, customs, values, that the individual overcomes their alienation from the objective world. Whereas the creation of a second nature in the subjective spirit was the transformation of the body and the development of particular aptitudes, facilities, and comportments, the cultivation of a second nature that Hegel describes in the *Philosophy of Right* is another way in which human beings are "born again" as social beings, transformed by inheriting and being inscribed by a nexus of discursive and ontoepistemic orders/norms.[31] Habit, in the context of thought and world, provides us with the infrastructure of our existence, creating "the embodied pathways in which and through we live our lives . . . [and] orientate ourselves in the world."[32]

It is here that habit's [*gewohnenheit*] important sense of meaning to dwell and dwelling [*wohnen*] should be kept in mind: habit concerns our inhabitation, the way in which we abide, our living in and sharing out of a world. The ideal version or operation of habit in the context of Hegel's sociopolitical thinking is that we can mutually constitute this having, that we can share out, or be at home with others in our fundamental having or being-at-home in the world. One manner of conceptualizing this could be, for lack of better terms, a kind of "communistic" habit: that we have or share worlds in which we can all mutually be at home. This is one reading of what Hegel means in his account of collective being-with or living with, what he calls *Sittlichkeit* or ethical life: we all share in the mutual formation, possession or having of the world; one is at home with the particular shape and distribution of the "ownership" or "property" of the world. If habit is at base a property, a sign of ownership, then what Hegel means is that in proper objective spirit, we all individually recognize ourselves and feel at home in the shape or form of the world. In Hegel, this is "spirit": we culturally, symbolically, and normatively feel at home in the sociosymbolic complexion and constitution of our world—the shape of the world we principally inhabit is one in which the material structures support us, prop us up, conforming to us as if to the "sole" of our feet. There should be, properly speaking, no privileged proprietors of habit—in this particular economy, there can be no split between the ownership and means of (re)production, but, rather, there should be participatory coproprietors who share mutually and reciprocally in sculpting the shape of the world. In this model, version of Hegel's dialectic of self and world in the *Philosophy of Right*, ethical substance—the large discursive, symbolic, normative field in and of which we are constituted—yields to our demands of its shortcomings when it does not accommodate us. The ways of thinking and being that structure ethical substance, for Hegel, comprise a living entity that is plastic, capable of being reformed and reshaped according to our requirements. When we do not feel at home in the world, Hegel stipulates that this substance is malleable and can and should be refashioned in order to fulfill and optimize our feeling of being-at-home, to adequately provide the conditions in which we can flourish. Moreover, Hegel maintains that this necessarily must be the case: objective spirit—community, society, and culture—is obligated, as per its own nature, its very purpose, to accommodate our not-being-at-home; the world is and should be for our living in it.[33] However, when the world does not reciprocate and reflect our needs, we are purportedly able to resist the shape of the world and reconstitute objective spirit so that once again the self and world reflect one another in mutual, reciprocal support so we can once again be "at-home."

HABITUÉS

Yet, this ideal account of the operation of habit becomes complicated read through the framework of Malabou's reading of habit. If we are to understand that habit functions homologously at the levels of both subjective spirit and objective spirit, then the specific accents of Malabou's reading which frame habit as the constitution of a proper "home" and a mark of possession, as something purposeful, amnesiac, and exclusionary, take on new resonances and consequences at the level of thought and world. If habit as an ontological mechanism is fundamental to the construction and maintenance, the stability and security of both the ways of being and thinking that structure the world, that give it form and allow it to reproduce itself, allowing one to be at home with and in the world, then it appears that there is a disavowed element to Hegel's presentation of habit that Malabou's analysis allows us to bring to light. The creation, maintenance, and reproduction of a purportedly "shared" ethical substance in which we can be at home through the work of habit is predicated and contingent on a founding, constitutive and ongoing ontological nonaccommodation: the constitution of a proper is always necessarily that of an improper, the inauguration of proprietors that of nonproprietors, a home and those who can inhabit and those who are not-at-home and cannot. To directly translate the accents of Malabou's reading at the level of subjective spirit to objective spirit, habit means: the creation of a home, a staking of possession and of property, that is concerted and intentional, that is forgotten by the habituated, those at-home, in the move from contingency to a second nature essentiality, and that necessarily entails that other habits—which means, at this scale, other ways of thinking and being, having and sharing out of and in the world—are foreclosed, made more difficult, improbable, if not impossible; or, in the starker language of Malabou's analysis, are effectively made to die: the living-on of a particular shape and form of the world is contingent on the preclusion or impossibility of others.

The seemingly small remark comprising Malabou's last words on habit in *The Future of Hegel*—that the future depends on habitués looking at their habits—here becomes decisive and presents a possible way to open up sociopolitical, ethical, and philosophical pathways to explore Malabou's thought. The habitués fleetingly mentioned are those *at-home*: they are the habituated and hence those that are at-one-with and at home in the world. The habitués can be characterized as those who contribute to the maintenance of the world—they are naturalists, conservationists, superintendents; they are *attached* to letting the world remain "naturally" as it is. This is because the habitués inherit and encounter a world in which they are welcome, accommodated because they in some sense comprise what is "proper" to the world, sharing in its possession or ownership: they are the patrons, the proprietors, the frequenters; they profit, invest in, *are invested in*, and subsidize the promulgation of the existing form of thought and world because it is *for them*. Yet, what of those who are *not* patrons or proprietors, but who are excluded from the accommodation of this "home," who are not integrated into the fabric of ethical substance, but rather exist as its constitutive refuse, its bystanders, the witnesses of the ostensibly "forgotten" history of the habitués and their world? Malabou's analysis then inadvertently discloses and forces us to articulate the existence of *nonhabitués*: those not-at-home, the improper, propertyless, nonaccommodated, the refused, the foreclosed. The nonhabitués are the excluded, the *expropriated*: if habit is the "sculptor" of plasticity, then the nonhabitués are those who have had their ontological right to shaping a world in which they can be at home with, which is the "proper," stolen, refused; their instruments blunted, chances taken, and possibilities muted. And it is this very expropriation, this impossibility, which acts as the ongoing possibility of the habitués world. The nonhabitués thus force a critical reappraisal of habit and plasticity: while the world is essentially capable of being changed or formed, an implicit requisite is that the world confronting the individual as an externality must actually be receptive to the individual(s) soliciting the change. Plasticity does not, as Malabou constantly avers, imply absolute flexibility: one does not merely dictate or simply enjoy the frictionless pliability of the world; the world is capable of resisting or refusing our efforts at transforming it. An underemphasized implication of this is that plasticity should be, perhaps, must be, thought in terms of who has access to its formative and formable potentials: that is, who marshals, enjoys, or "owns" the efficacy of plasticity; that is, if habit is again essentially the sculptor of the plasticity of the world, the possibility that it can shaped, formed, and transformed, if it can take any shape, anything that we put it into, then who and what "we" wields the chisel? The habitués are this "we." If plasticity entails the ability to resist the donation or imposition of form, one has to think that this capacity is not uniformly deployed in a transformative capacity, but can be equally utilized in a conservative one, that is, those who are attached to and are in the interest of preserving their own shape of life—the habitués—against the incursions of others who would imperil the comfort of their "home," their possessive, privileged structure of life: the nonhabitués, the expropriated.

It is this refusal that the habitué does not see, but the nonhabitué does: the one who feels at-home in and at-one-with the world experiences an unperturbed carrying-on. The habitué does not experience or countenance that the condition of possibility of their world, the one which is made for them, is precisely the condition of impossibility for others: that the present form of the world was and is not made for them. This is because habit operates and earns its recalcitrant efficiency through "self-forgetting," that maintains its efficacy and power through the continued patronage of amnesiac habitués, who function as preservationists of an apparently ongoing-present in which a past, a history that constituted this present does not obtain, matter, or perhaps even exist. Yet, there are those who do not forget, who are not permitted the comfort that accompanies the criminal amnesia of indifference wrought by the work of habit, but bear the burden of witnessing and suffer the complicit and violent obliviousness of the habitués. If the proprietors of the world, the ones who are at home in and one with the world, are those who reiterate and reproduce the same through the self-forgetting work of habit, who invest in, patronize, and subsidize the world as it is at the expense of what it could be, then the chance of lucidity, remembrance, and a "seeing-clearly" are not to be found here, but elsewhere. The possibility of thinking and living differently, of awakening a vital energy that could transfigure the form of the world, lies not within the patrons of the familiar, but those that have been refused and can clearly see the familiar for what it is.

What are we to make of Malabou's prescriptions then regarding habit? With respect to the notion that philosophy can spark off a "necessary anamnesis" that can desediment the inertia of existing habits of thought, is philosophy itself not part of this intricate production? It is, as Malabou writes, "a woman's tomb," and, not only that, but has acted as the historical "proper" of thought that has provided the intellectual infrastructure of and for the habitués that structure the modern world: antiblackness, racism, colonialism, sexism, ableism, white supremacy, and capitalism.[34] It then seems incumbent on philosophy to detach from itself and let it be informed by its own history, by that which it is has othered to constitute itself. Disciplinarily, then, and ontoepistemically, these intellectual nonhabitués, other ways of living, being, and thinking solicit and force philosophy to remember its past and hold it to account; the habitués or patrons may not remember, but there are those that do. The sovereignty of philosophy is a death promulgating chimera, one that must be disabused. If we follow the interpretive path elaborated here, and its interplay and implication of those who forget and enjoy, and those who remember and suffer, this seems to point to the indispensability of other onto-epistemic practices to radically alter and transform the pathway and future of philosophy.

And what of the habitués and the call to look at their habits? This becomes an imperative task of questioning ourselves to what are we blinded and what do we assist in and abide by the cataracts of habit? If living in the world is a dwelling, by means of habit, and this means an abiding, our existence is constitutively intricated with the question of what do we abide by in existing: what do we invest in, patron, permit simply by continuing on in the forms of life that we inherit and call home? The question of habit thus evokes the question of if we are capable of detaching from our current habits of thinking and forms of life to make space and give way to other forms, impressed and guided by the hands, voices, and visions of others, that are not the ones that we are historically attached to and seek to preserve. We are not only capable of such a task, but we must, lest we look into the face of others as the guarantors and executioners of their disaffection, disenfranchisement, and death. If we are not discomfited by this station, we remain attached; if we respond, find ourselves responsible and accountable, obligated to account, we must detach and become the material by which other's lives are formed for the better—not in spite of, but most of all and essentially, at the expense of the present shape of our own. Are we plastic enough to risk, refuse, and resist our own attachments to a form of the world in which we are at home, but whose condition of possibility consists in the not-being-at-home of others? The promise of plasticity is that it is surprising enough that it may be possible to radically transform the structure of the world, that the "universal" is truly universal enough to accommodate each particular; that the world itself can bear and take any form or shape. Such is the plastic passion of the world: that it can suffer, it can endure any metamorphosis. Such a radical, utopic thrust animates plasticity and gives it its hope. But, again, what of the history attested to by the nonhabitués, the expropriated, who have suffered the habitués world, its recalcitrance, its obstinacy, and its stupid cruelty and purposeful violence? Here emerges an urgent impasse confronting Malabou's philosophy: the transformability and metamorphic possibility of every structure; the inveterate rigidity of the exclusionary ontological structure of the world.

NOTES

1. This research was supported by the Social Sciences and Humanities Research Council of Canada.

2. G. W. F. Hegel, *Philosophy of Mind*, trans. William Wallace (New York: Oxford University Press, 1971), §391.

3. Ibid., §404.

4. Ibid., §403.

5. Ibid., §405.

6. Ibid., §407.

7. Ibid., §410.

8. Ibid.

9. Ibid.

10. Ibid.

11. Ibid.

12. Catherine Malabou, *The Future of Hegel Plasticity, Temporality and Dialectic*, trans. Lisbeth During (London: Routledge, 2005), 26.

13. Ibid., 37. This description resonates with Malabou's recent work on the transcendental, which she also describes as a mark of property, a kind of possession. What is the relation between habit and the transcendental? Does habit effectively function as a kind of transcendental, a condition of possibility? See Malabou, "Can We Relinquish the Transcendental?"

14. Ibid.

15. Ibid., 68.

16. "Existence reveals itself as plasticity, as the very material of presence, as marble is the material of sculpture. It is capable of receiving any kind of form, but it also has the power to give form itself. Being the stuff of things, it has the power to both shape and to dissolve." Catherine Malabou, *Plasticity at the Dusk of Writing*, trans. Carolyn Shread (New York: Columbia University Press, 2010), 81.

17. Hegel, *Philosophy of Mind,* §410.

18. Malabou, *The Future of Hegel*, 74.

19. Ibid., 75.

20. Ibid., 66.

21. Ibid., 76.

22. Ibid.

23. Ibid., 190.

24. Ibid. Emphasis mine.

25. Hegel, *Philosophy of Mind*, §410.

26. G. W. F. Hegel, *Philosophy of Right*, trans. Alan White (Newburyport, MA: Focus Publishing, 2002), §151z.

27. Simon Lumsden, "Habit, *Sittlichkeit* and Second Nature," *Critical Horizons* 13, no. 2 (2012): 238–39. For further discussion of the role of habit in different aspects of Hegel's thought, see also Simon Lumsden's "Second Nature and Historical Change in Hegel's Philosophy of History," *International Journal of Philosophical Studies* 24, no. 1 (2016): 74–94; "Habit and the Limits of the Autonomous Subject," *Body & Society* 19, no. 2 and 3 (2013): 58–82; "At Home with Hegel and Heidegger," *Philosophy Today* 59, no. 1 (2015): 7–21.

28. Hegel, *Philosophy of Right*, §145.

29. Ibid., §146.

30. This experience of alienation from the world is effectively a variation of what Hegel characterizes as the Unhappy Consciousness, a state in where the individual perceives itself to be changeable and unessential in contradistinction to the objective world which is essential and unchangeable. Hegel, *Phenomenology of Spirit*, §208–209.

31. Hegel, *Philosophy of Right*, §151.

32. Lumsden, "Habit, *Sittlichkeit* and Second Nature," 226.

33. "There is nothing degrading in being alive, and there is no higher spirituality within which one could have worldly being. It is only the raising of the given to something self-created that yields the higher orbit of the good." Hegel, *Philosophy of Right*, §123.

34. Catherine Malabou, *Changing Difference: The Feminine and the Question of Philosophy*, trans. Carolyn Shread (Malden, MA: Polity Press, 2011), 100.

BIBLIOGRAPHY

Hegel, G. W. F. *Phenomenology of Spirit*. Translated by A. V. Miller. New York: Oxford University Press, 1977.

———. *Philosophy of Mind*. Translated by William Wallace. New York: Oxford University Press, 1971.

———. *The Philosophy of Right*. Translated by Alan White. Newburyport, MA: Focus Publishing, 2002.

Lumsden, Simon. "Habit, *Sittlichkeit* and Second Nature." *Critical Horizons* 13, no. 2 (2012): 238–39.

Malabou, Catherine. *Changing Difference: The Feminine and the Question of Philosophy*. Translated by Carolyn Shread. Malden, MA: Polity Press, 2011.

———. *Plasticity at the Dusk of Writing: Dialectic, Destruction, Deconstruction*. Translated by Carolyn Shread. New York: Columbia University Press, 2010.

———. *The Future of Hegel: Plasticity, Temporality and Dialectic*. Translated by Lisbeth During. New York: Routledge, 2004.

Section 3

TOWARD A PASSIONATE PHILOSOPHY

Part I

THE LIFE OF SCIENCE

Chapter 8

After Deconstruction? The Challenge of Malabou's Plastic Biohistory

Joshua Schuster

How can one respond to the call Catherine Malabou makes in her book *Changing Difference* (*Changer de différence*, 2009) that we need to find a way to create concepts and pose problems that are not already covered by deconstruction? Malabou writes about the need to elaborate new ways of thinking and new figures that are not necessarily inherited from Derrida and his own philosophical preferences. Malabou points to two forms of leaving deconstruction behind in her work. First, Malabou asserts a break from deconstruction in the activity of being a female philosopher, embodying the changeable condition of the feminine, without reducing the feminine to a static biological essence but also without consigning the feminine to being nothing but a performative difference. Second, Malabou writes of her discovery and elaboration of plasticity as "metamorphosis"[1] and a metabolic difference, defined as the capacity for an object or concept both to give and receive form. Plasticity is not reducible to the deconstructive logic of the double bind or an assertion of the ultimate textuality of the world. Instead, plasticity recalls the original malleability of form and essence, not as the indefinite play of deferral and not as the preformed dialectic, but as a property of transformability that is already inherent in form itself.

Malabou claims that we are no longer solely in the epoch of deconstruction, defined as an epoch of writing always poised between the continually shifting modalities of presence and absence. Rather Malabou asserts our epoch is better described as one of plastic ontologies and partially modifiable histories and forms. Malabou finds that Derrida himself began to intuit a need to move beyond deconstruction with his brief forays into thinking the undeconstructible. I will quote Malabou at length:

> In some ways Derrida himself gradually abandoned, or at least transformed, the imperative to deconstruct presence. . . . As proof, I'll mention the emergence of

> the problem of the "undeconstructible," that Derrida outlined in his late work under the names "justice" or "democracy." If there are un-deconstructible cases then that means they can come back, that in a certain sense they cross the two rings of fire that are the history of metaphysics and the era of deconstruction. The un-deconstructible is not of the order of presence, but it is just as much a form of resistance to the text. Derrida therefore had to admit that there was a form of substance that, without being a parousia, can no longer be confused with the incessant mobility of graphic difference. It appears, however, that for lack of time he did not sufficiently examine the ontological consistency of the un-deconstructible. . . .
>
> To my mind, this is the essential question Derrida bequeaths us.[2]

The cases of what could resist deconstruction are plural. Here I want to articulate a number of them:

1. What cannot be deconstructed is the transcendental. The transcendental, if it exists, is independent or unconditioned by time and space—and anything in space and time can be deconstructed. Anything in time and space can be separated from itself, negated, fail to be reiterated, overcome, or dispersed. However, if there are any transcendental undeconstructibles, they must give an account for their being in the world, and thus are capable of being deconstructed in that realm.
2. What cannot be deconstructed can be said to be self-causing and self-sufficient, whether as an essence or as a conceptual category. Yet Derrida repeatedly argued that every essence and concept requires a supplement or difference that it cannot avow but contributes to the conditions of existence for any essence. No form of being and no conceptual category can exist as closed and self-sufficient since beings and concepts exist as differentially related to other beings and concepts.

   However, as Malabou points out, sometimes Derrida himself made questionable assertions of localized categorical closure and fixity, and she cites, for example, Derrida's assumption that neurobiology is fixed like a computer program is fixed. For a different example, I would also point out in another brief remark that Derrida once asserted that eating and consumption necessarily defined the living subject, which prompted Derrida to declare, "I do not believe in the noncarnivore."[3] But there is no reason to define consumption as such as carnivorism, because in addition to the entire world of herbivorous animals, there is no basis for saying that carnivorism or even the broader notions of eating as appropriative activity this term implies describes the nutritional activity/passivity of plants. So these kinds of local and specific categorical undeconstructibles in Derrida's own work can and should be deconstructed.

3. What cannot be deconstructed is what cannot be foreseen in advance because it is always futural. The futural will always surprise, will always take the form of a promise, and will never fully arrive in the present. The problem here is that the future may not always arrive, or at least arrive in this form defined by the promise, because there is no indefinite subject, no infinite promise for the subject, so there is a way for this kind of futurity to not apply to possible futures.
4. What cannot be deconstructed is what is declared to be a universally applicable good, defined as either a formal or a practical principle. Derrida never asserted any blanket statement of ethical principles that could be absolutely and irrefutably reasoned in advance. Rather, ethics only begins in an encounter, an interruption, a surprise, which reveals all along the capacity for unconditional openness built into the ontological conditions of living beings. This unconditional openness cannot be deconstructed in advance, but every actual encounter with otherness effectuates an ongoing series of responsibilities, deferrals, and asymmetrical demands that are open to deconstruction.
5. What cannot be deconstructed is what is singular. What is singular is irreducible, incomparable, unique, and refuses to be systematized. However, the singular is only known because it is inscribable and comparable, because it is rereadable, reiterable, and reinterpreted into a system of meaning and experience. But the singular is not the same as the repeatable. The singular, without double binds yet also not recognizable outside of the double bind, is what resists deconstruction and what exemplifies it. Much of Derrida's later work is focused on arguments for singularity rather than play or continual difference. For Derrida, the singular must be broken but still remain singular. If there were only singularities, nothing would be thinkable since there would be no way for thought to access the singular; however, at the same time, every subject lives a unique and singular world. Every living being is irreplaceable and has a unique perspective within the world. But every living being must break with this singular perspective and encounter the constant otherness of the world in order to survive. Every subject must both retain and forfeit its singularity at every moment, which is to say, both embrace deconstruction and reject deconstruction.

This is an exploratory list and I am sure there is more to add. What immediately stands out is how the resistances to deconstruction come primarily from the extremes of the philosophical field. The most general, the universal and the transcendental, can exceed deconstruction, as well as the most specific, the singular, can refuse deconstructive inquiry. The upper limit and lower limit of thought and being haunt deconstruction and are haunted by deconstruction in turn.

This list of undeconstructibles can help clarify what Derrida gestured to in his later work regarding the limits of deconstruction. In an interview in 1994, Derrida asserted, "If anything is undeconstructible, it is justice. The law is deconstructible, fortunately: it is infinitely perfectible. I am tempted to regard justice as the best word, today, for what refuses to yield to deconstruction, that is to say for what sets deconstruction in motion, what justifies it. It is an affirmative experience of the coming of the other as other."[4] The affirmation of an unconditional openness to this otherness cannot be systematized but also cannot be deferred indefinitely, since the other can always arrive at any time.

This example of justice outside of deconstruction constitutes the firmament of ethics, in that to be a subject is to be always potentially exposed to and interrupted by an other, necessitating some kind of responsiveness to that other. Derrida does seem to imply in much of his discussions of responsibility a latent reference to biological notions of the capacity to affect and be affected, although Derrida certainly would not point to a biological basis for his ideas. Derrida views exposure, interrupt-ability, and response-ability toward the other as ontological conditions of being a subject. But biological questions are still germane to these ontological claims, and there are two clear possibilities where this kind of ontological view of justice falters at biological levels: (1) there can be cases where there is no other, such as in the event of extinction; (2) more broadly, we can question how interrupt-ability, where the other makes a kind of cut, difference, or disruption into the self, can account for how biological forms develop these affective capacities to begin with. We need to ask what are the conditions of biological form such that a cut can be made into them in the first place? But also, what kinds of cuts allow for further cuts to be made, and what kind of cuts are terminal or terminating? Malabou's propositions on the plasticity of biological forms and the metamorphic potentials of "biohistorical" bodies can require a wider conceptual and ontological thinking than notions of inscription, difference, and responsibility that are fundamental to Derrida. A significant portion of the challenge Malabou has posed to Derrida has circulated around the conceptual stakes of contemporary biological thought. I turn now to a closer look at some key aspects of Malabou's claim for "changing difference," and how shifting from Derrida's scene of writing to matters of biological form raise a whole new series of potentials and problems for contemporary thought.

THE ENDS OF DECONSTRUCTION

Looking again at the cases of what is undeconstructible for Derrida, one finds that so much of what is said to be beyond deconstruction relies on the continuation of a living subject, if not a wholly human subject, who finds him or herself embedded in a partially legible world whose continued legibility is taken for granted. If Derrida finds that justice, democracy, hospitality, and a "messianism without messianicity" seemed to be beyond the deconstructive work of the trace that both inscribes and effaces, these apparently structural conditions only exist for a living subject who can be continually surprised by an other. Each of Derrida's supposedly undeconstructibles appear tied to a subject who is not at all transcendental, but rather dependent on temporary configurations of life on an unstable planet in a universe that is only temporarily hospitable to life.

Furthermore, we also live in a time where we are witness to an increasing rate of the disappearance of life on earth. Extinction wholly eliminates the futural conditions for a species. Every extinction is a singular incomparable loss that takes away a form of singularity from the Earth. Yet this destruction of the singular species is not the same of the breaking of the singular subject in deconstruction. Derrida's broken singularities are the conditions of possibility for subjectivity in the first place, while extinction is the erasure of possibility for a subject. Derrida has repeatedly sought to deconstruct concepts of ends and finitude in philosophy, beginning with his essay "The Ends of Man" (1968) in which he shows how a plurality of concepts of human ends and finitudes are associated with metaphysical claims for humanity's *telos*. But extinction is a definite end, not a pluralism of ends, and the radical finality of any life, what Derrida later calls "the end of the world" in any person's death, poses an irreversible end that provides no clearly meaningful *telos*. With these cases in mind, Malabou is correct to call for a changing thought that revaluates the concept of difference for our time. Moreover, in the time of the Anthropocene, the new name for our geological period that is defined by humans reshaping earth processes and the reshaping of the human by these same processes, we are becoming aware of past Earths and possible future Earths that are not hospitable to the conditions of life that have generated the conditions for hospitality in the first place. Thus justice, hospitality, and unconditional receptivity to the other are themselves conditioned by a certain configuration of life and planet that are clearly temporary. The changing Earth will move beyond deconstruction in one way or another. There are futures without hospitality.

But can we imagine the opposite—a world where deconstruction would always exist? This to me seems impossible. If deconstruction as a means of conceptualizing the world (rather than just describing the mechanical work of spacing and timing) were always at work, infinitely and indefinitely, it would be guaranteed and thus fixed. Deconstruction then would be a system, always running and always operational. It would be like a Hegelian process that would never fail. But deconstruction cannot ever be this system, backed by a set of guarantees, wholly immune to being inoperable. Thus what is undeconstrictible is the very notion that there is no permanence to deconstruction. We can never rule out that deconstruction might be illegible or unworkable at any moment. If deconstruction always has a future, there would be no more promise, there would be a guarantee, a fixed, secure structure. Deconstruction, to exist at all, then must always be open to failure and finitude, including the failure and finitude of its own method and actualization.

This articulation of a world without deconstruction might seem bleak, and it might seem like the kind of world we should do everything we can to prevent, even if we sense that there is a degree of inevitability and admit that deconstruction could never be perpetual. However, what I want to argue here is for a continuation of Derrida and Malabou's arguments for the need to begin to imagine what is beyond deconstruction in order to understand how it is possible that deconstruction might be a momentary meeting of earth and world, to use Heidegger's terms. In particular, I want to follow Malabou's direction in looking further at biological conditions that are not wholly defined by the trace structure posited by Derrida. Unlike Malabou, however, I find myself more attracted to thinking biological problems not on the scale of plasticity of DNA, stem cells, or neurons, but rather on the scale of species and planetary ecological conditions. Darwin's combination of empirical and theoretical work provides an especially potent resource for such inquiries, especially as many readers of Darwin have shown that his research shares many concerns with deconstruction and continental philosophies of life. Darwin is not just a theorist of the mechanisms that make life but he also keenly examines the mechanisms that undo life. Furthermore, I have found Darwin crucial not just for theorizing life and extinction, but to also to make sense of ecology, the Anthropocene, and recent models of biohistory. It is a puzzling omission that Darwin's ideas are not much engaged by Derrida, and neither by almost any other French continental philosopher. Just up until recently, Malabou had not sought to draw much from Darwin's work either. In considering Darwin's biohistorical thought, philosophy should not be grounded in biology, but it can be problematized by a wide range of biological events and concerns.

Although Derrida claimed his philosophy had long been concerned about the "animal of reading and rewriting,"[5] he rarely engaged directly with biological or ethological thought and research. In his lectures on *The Beast and the Sovereign*, Derrida argues that deconstruction should reject any reliance on "commonly accredited oppositional limits between what is called nature and culture, nature/law, *physis/nomos*, God, man, and animal or concerning what is 'proper to man' than to muddle everything and rush, by analogism, toward resemblances and identities. Every time one puts an oppositional limit in question, far from concluding there is identity, we must on the contrary multiply attention to differences, refine the analysis in a restructured field."[6] Deconstruction works to multiply the differences, which include both overlaps and exclusions, between animality and humanity. As powerful as this approach can be, I want to argue here that multiplying differences can be only one tool to understand the biological conditions that make and unmake species and ecosystems. Extinction and biodiversity, the twin concerns of this chapter, are not primarily problems of understanding the multiple differential nexus of animality and humanity. Extinction and biodiversity intertwine with each other and affect all life but are not reducible to "biologistic continuism,"[7] which Derrida rejected for its simplistic, single explanatory position. Following Malabou's guidance, we must consider then how an analysis of biological forms and the biological undoing of these forms will require a combination of tools within and without deconstruction. Following Malabou, we must inquire into what constitutes the subject of biohistory who is increasingly aware of how extinction as well as biological plasticity and the plurality of biological forms are interconnected.

SPECIES FORM AFTER DECONSTRUCTION

Very early in his thinking, Darwin understood that any theoretical model of biology had to be able to account for speciation as well as the loss of species. While still traveling on the *Beagle*, Darwin wrote in 1844, "Certainly, no fact in the long history of the world is so startling as the wide and repeated extermination of its inhabitants."[8] By the time of *On the Origin of Species* (1859), Darwin came to regard extinction and speciation as fundamentally intertwined, such that the demise of some species occurred precisely because other species evolved in a way to secure the conditions of life at the expense of other species. To think speciation, then, one had also to think extinction. Extinction is built into the process of speciation, not a deviation from it, yet extinction can also dissolve the process of speciation altogether.

Darwin developed a theory of biology to account for the empirical end of species, but found that he did not need to have a strict definition of the difference between species and variation, or the difference between individual and species group. "I look at the term species, as one arbitrarily given for the sake of convenience to a set of individuals closely resembling each other, and that it does not essentially differ from the term variety, which is given to less distinct and more fluctuating forms. The term variety, again, in comparison with mere individual differences, is also applied arbitrarily, and for mere convenience sake."[9] He also did not or maybe could not provide a strict, watertight definition of life. This cautiousness on Darwin's behalf is humbling. Darwin did not need a metaphysical definition of life affixed to first causes or absolute principles. Rather he only sought to describe the physical conditions of living organisms in action, the constraints and opportunities felt by all forms of life. It is still cautioning to admit that even today biologists and philosophers do not have a single cohesive definition of life. Noam Lahav lists forty-eight definitions of life that have been proposed in the past two centuries.[10] There is room for even more, as one of the conditions of life is the potential capacity for organisms to change the conditions, material properties, and forms of life, thus changing the definition of life. In not defining life according to any metaphysical programs, Darwin does not dismiss theories of life but actually encourages more of them. To think life one must think how such thought could be revisable as the conditions of life change. The provisional definitions of life then follow from the very activities of life in participating in the making of conditions of life. Later I will propose a similar argument for how biodiversity can also only be provisionally defined and must be continually revised by the processes of biological life.

Darwin's strategically "weak" definitions of life and species are knowingly both vaguely indeterminate and full of nuance. Darwin still keeps the species form as a key reference throughout *On the Origin of Species*, even as he advocates for its deconstruction. The notion that one could both deconstruct a category yet still retain a conceptual coherence to the category aligns with recent arguments for "postmetaphysical" forms of thought that, according to Michael Marder, assess "the material, the singular, the finite"[11] while trying to do justice to what metaphysics has suppressed. Marder writes that postmetaphysical thinking, in the context of theorizing the heterogeneity of life, consents "to let beings be, to save singularities from the clasp of generalizing abstraction, and perhaps to put thought in the service of finite life" (5).

One must explicitly contrast this approach with any version of generic vitalism that asserts a metaphysical impetus or power for life beyond any empirical instantiation of living beings. Deleuze's vitalism is a quintessential example of elevating metaphysical powers over empirical, physical conditions of existence, to the extent that he conflates life and metaphysics. In one of his last essays, Deleuze equates the movement of immanence with the existence of what he calls "a life."[12] "A life is the immanence of immanence, absolute immanence: it is complete power, complete bliss" (27). This "a life" is "indefinite life" (29), and no form that a specific life takes can ever have any lasting effect on the limitless and indefinite immanence of life. For Deleuze, the entire universe is comprised of qualitative and sensual acts of becoming that are perpetual and inexhaustible. The species form of life is just one version of how this "a life" can be actualized. The death of a species, or any extinction of life, is just a local stoppage in a metaphysical process that is preprogrammed for becoming and continual individuation. Ultimately nothing of the physical properties of life can have a permanent and lasting effect on the metaphysical properties of "a life," which he states is a transcendentally empirical power.

For Deleuze, extinction is no problem physically or philosophically. In stark contrast, Ray Brassier argues that extinction is a "transcendental trauma"[13] that ruins any metaphysical, transhistorical claims for consciousness, life, or meaning-making beings. Brassier analyzes extinction's devastating impact on philosophical claims for the special relation between life and meaning. The upshot of Brassier's argument is to call for a shift from continental life philosophies to naturalist and rationalist theories of mind. Surprisingly, he has little to say about the biology and ecology of extinction. In these two polar opposite approaches, extinction is either irrelevant or it is so dominant that it forecloses any further understanding of the nuances extinction can play in speciation, including the way that, as Darwin discerned, extinction can play a creative role in provoking further speciation. Assuming that extinction makes all thought of life moot in the last instance ends up providing little insight into the biological and philosophical issues of living with particular cases of extinction and coming to know the specific processes and limits of life. A self-reflective thought on extinction must inquire not just into limits of life and thought, but also entails cultivating forms of knowledge and ethics appropriate to sharing a planet teeming with species for now. This is not to dismiss abstraction in philosophical thinking about life, but rather to argue for further understanding how the tenuous connection between thought and life overlaps with the time and form of species. Neither Deleuze nor Brassier delves into the specific affordances and failures faced by creatures in their worldly ecologies.

DARWIN'S PLASTIC PAST

While most of Malabou's work to date on the theorization of biological concepts focuses on cellular and specifically neural plasticity, she has recently given more attention to issues of biopolitics and the biological sciences more generally. In a newer essay, Malabou examines Darwin's use and implication of the term "plastic" in *On the Origin of Species*.[14] Malabou states that "plasticity constitutes one of the central motifs of Darwin's thought" (50). Malabou adds: "Plasticity situates itself effectively at the heart of the theory of evolution" and prompts the question, "How does a 'form' take form according to Darwin?" (50). At times, Darwin does indeed link concepts of plasticity to biological form; however, Darwin employs the term "plastic" just four times in his book. Malabou comprehends Darwin's usage of the concept of plasticity as a synonym for the evolutionary processes of variability of any organism. While I think this interpretation is ultimately justified,

Darwin has a more constrained use of "plastic" and technically only uses the term to apply to variability that is induced by domestic breeding. Each of the four times Darwin uses the term "plastic" comes within the context of a discussion of artificial selection or domestication, for example, in this quote that Malabou cites: "Under domestication, it may truly be said that the whole organisation becomes in some degree plastic."[15] Though Malabou finds this quote echoes her own expansive thinking of ontological plasticity, specifically Darwin has in mind an association between the plastic arts, such as sculpture, which the artist can manipulate at will, and domestic breeding practices that have artificially selected for particular traits and produced, for example, over 300 different dog breeds.

Malabou generalizes Darwin's usage of plastic to refer to any variation produced in natural or artificial circumstances. "Characteristic of variability, plasticity designates the quasi-infinite possibility of changes of structure authorized by the living structure itself,"[16] Malabou states. I think Malabou is right to extend the concept of plasticity across natural and artificial selection, as Darwin himself never makes a strict divide between the two processes. To support the notion that Darwin theorizes an overlapping of artificial and natural selection, Darwin begins *On the Origin of Species* with several chapters on breeding and domestication because it is easy to see the role breeders play in changing a species' characteristics within a few generations. Yet there are also reasons to demure from Malabou's extending of plasticity to a "quasi-infinite of possibility of changes of structure." The plasticity of the organism in domestication is due to conscious, goal-directed, teleological design, and controlled efforts at selection, in contrast to natural selection that is unconscious and has no overall design, guiding hand, or teleological trajectory. Yet in both cases, variability occurs only based on slight differences from previous biological forms, and plasticity is restricted to variations on existing templates.

There is a more consistent term that Darwin uses to describe the changing variations across species, whether natural or artificial: modification. "Modification" or "modify" is used 405 times by Darwin in *On the Origin of Species*. It is his preferred term to describe how the plasticity of an organism is tied to the alteration/modification of a previous species form. Modification occurs when the descendant of an organism includes a physical alteration, often slight, from its parent generation. Malabou often uses modification as a synonym for plasticity, but in Darwin's case there is a slight difference between the terms. Modification is not best defined by Malabou's claim for "quasi-infinite possibility" within biology—rather modification emphasizes that any biological change has to stem from a preceding organic form that also has the possibility to give rise to variation and allow for that variation to be transmitted to subsequent generations.

Another way to comprehend the importance of modification is to grasp how biological plasticity is always constrained as if passing through a bottleneck of formal possibilities. A bottleneck is some form of material constraint that both encourages and delimits plastic possibilities. The species form, its genotype and phenotype, acts as a bottleneck such that new forms can only be variations based on previous templates. Thus, instead of a following a preset principle of quasi-infinite becoming, plasticity, difference, or metamorphosis, speciation occurs by modification, typically incremental, of a previous substrate. Forms must build on and from previous forms, which makes the existing, available diversity of species forms extremely important for producing subsequent life. Plasticity is constrained by previous plastic forms, and perhaps paradoxically, this constraint sets the stage for the multiplicity of biological difference. Even if the whole of an organism can be said to become plastic via domestication (now grouped under the broader work of biotechnology), plasticity does not mean that any aspect of life can be modified at any time. The role of biological bottlenecks and the restriction of modification to only already existing organismic forms is a major reason why the pressure of extinction bears much more significance on the diversification of life than is credited by vitalist philosophies.

We cannot understand the continuous plasticity of life without connecting organismic change to the bottleneck of previous species forms and to the failure of plasticity in extinction. Stephen Jay Gould argues provocatively that, based on the evidence uncovered at the Burgess Shale, there was much more dramatic diversity of species body plans 530 million years ago than today. Gould well acknowledges that the total number of species and subspecies is much larger in the current age. What Gould points to, with the Burgess Shale as his evidence, is how the relatively early explosion of multicellular anatomical plans provided a much wider and disparate template for species forms and modifications compared to today.

> The sweep of anatomical variety reached a maximum right after the initial diversification of multicellular animals. The later history of life proceeded by elimination, not expansion. The current earth may hold more species than ever before, but most are iterations upon a few basic anatomical designs. (Taxonomists have described more than a half million species of beetles, but nearly all are minimally altered Xeroxes of a single ground plan.) In fact, the probable increase in number of species through time merely underscores the puzzle and paradox. Compared with the Burgess seas, today's oceans contain many more species based upon many fewer anatomical plans.[17]

Malabou is right to indicate that any aspect of any generation can be modified or made plastic, but there are not infinite (or "quasi-infinite templates for becoming and difference. There have been times in the history of life on this planet when species forms were much more distinct from each other. The rates and intensities of biological plasticity can fluctuate dramatically as the number and disparity of species forms fluctuates.

BIOPOLITICS AND SPECIES AGENCY

Malabou's reexamination of Darwinian plasticity connects to her recent critical analysis of Foucault's elaboration of biopolitics. Malabou finds in Foucault a theorization of biology as always being captured by political power. For Foucault, bodies and their biological conditions are passive in the face of historically constructed categories, social structures, institutions, and dispositifs that seek to actively shape life to political ends. Foucault certainly does theorize a counterpolitics of bodies, especially at the level of the pleasures and cultivations of the self, but he does not consider how life already poses, in its biological activity, its own resistances and creative responses to the biopolitical manipulations that subject life. Malabou claims that one cannot find in Foucault, or in any other developments of biopolitics, including the work of Agamben and Esposito, a theoretical appreciation of the agential, activist qualities of biological bodies, a "*bio*logical resistance to *bio*power."[18] Malabou goes on to argue for a refusal to separate biology from what she calls "the symbolic"—any cultural category said to exist in opposition to biological nature, such as spirit, or conceptual structures that are said to be founded only self-differentially and without any ties to physical conditions. Malabou also protests against separating biology and the sphere of political activity (which Hannah Arendt so strenuously sought to cordon off from being based in biological activity), although she also questions the conflation of the two with the term "biopolitics." Instead of assuming that biology is passive in the face of biopolitical control, or that biology is based on a preset naturalist program impervious to social circumstances, Malabou calls on philosophers to finally address the overwhelming research in contemporary biological sciences that describe the agential lives of species and the imbrication of social and natural activities among species.

This call to consider agency and plasticity as constitutive of biological forms themselves, however correct, still must be based on recognition that the existing diversity of biological forms and ecological conditions provides both constraint and opportunity for these modifications. Plasticity and agency must always riff off of previously existing living templates and environmental circumstances. The disciplinary tendencies of biopolitics perhaps then can be counterbalanced by the creative potentiality within biological life; however, there is no generic creative plasticity, transhistorical vibrant matter, or automatic differential work of possibility in biology. Furthermore, the plasticity of life occurs within the context of aspects of species sexuality that can exceed mere survival and reproduction, as well as the potential for the loss of species forms in extinction. Sensuality and precarity, overproduction and decimation, are themselves imbricated in the various plastic templates of life.

Darwin came to argue that speciation was always ongoing, up to the point of extinction. In a sense then, species were most clearly definable only at the point that they no longer existed. Once extinct, one could pinpoint what a species specifically and finally entailed, the only point where species difference became fixed and irreversible. In extinction, essence appears only after existence. Paradoxically, when a species loses its form, this is also the situation where form is finally fixed for that species (even if a species can be revived or deextinct through technology, the revived species would carry a differential mark of human creation and would not be the exact continuation of a previous form). Every species then has this double essence, one essence aleatory and plastic, terms Malabou often uses together, and one essence that appears only when the species disappears. What then constitutes a species' horizon is both plasticity and finitude, ongoing modification and irreversible disoperation.

A remarkable example of the coexistence of these seemingly contradictory states can be found in the situation of coral reefs. Coral is comprised of the exoskeleton made from the calcium carbonate secretions of numerous polyps that cluster in colonies. The slow build-up of calcium carbonate turns into the rocky base upon which further growth by polyps can take place. The multicolor appearance of coral is due to algae that grows symbiotically with the coral. Coral are thus "part animal, part vegetable, and part mineral, at once teeming with life and, at the same time, mostly dead."[19] It is estimated that hundreds of thousands of other aquatic species depend directly and indirectly on the life that congregates in and around coral reefs. There is plenty of evidence that corals are both powerfully resilient and extremely fragile beings. Coral reefs can be damaged by storms and are vulnerable to chemicals dumped in waters as well as overfishing and breakage from human contact. Most dire, however, is that corals are extremely sensitive to global warming. Corals need warm water with a finely delimited range of pH solution. The increasing saturation of carbon dioxide in the air falls into the ocean and increases the acidic level of the waters, which is disrupting the reproductive cycles of coral. The acidification level of the oceans has changed over the Earth's history, and corals have come and gone with these variations. Different reef-making organisms have proliferated and gone extinct in the past, especially during mass extinction events. There are even several gaps in the biological record that show no coral for several million years in each gap. Over the past 500 million years, different species have evolved to make corals while other coral-making animals have gone extinct. Corals are a mix of the living and the dead that have been and continue to be enablers of zones rich in species diversity. Yet the dying of coral today is also one of the prime indexes of the collapse of species numbers in the mass extinction event that is one of the markers of the Anthropocene. Corals exist in the mediation between biology and geology, life and death, extinction and proliferation, animal and mineral, plasticity and precarity, all while harboring much of the ocean's biodiversity.

PLASTIC BIODIVERSITY

I have argued here that Malabou's theorization of plasticity and biopower can help to shift philosophical thought on life away from unlimited and undeconstructible difference (Derrida) or endless becoming (Deleuze). Contra Bergson, there is no guarantee of creativity in evolution. Plasticity is best contextualized within biological science's empirical and theoretical network that accounts for how the proliferation of species forms is imbricated with the loss of species forms. As Malabou states, "Plasticity is not, I repeat, an empty transcendental instance. Plasticity is nothing outside its context and its supplementary status. The modifiability of concepts does not exist outside the specific historic modes of their modifications."[20] In the case of biological plasticity, modifications must follow from variations on existing forms. The accumulation and multiplication of platforms for plastic modification can be described as a process of biodiversification. In closing, I want to briefly examine how the concept of biodiversity can lead to a further philosophical intensification of biology that can continue shifting the theorizing of life away from templates of metaphysics and deconstruction, even as deconstruction's tools remain helpful.

The term "biodiversity" was coined only just recently in the 1980s, and, just like with the difficulty of defining species, there is still no exact agreed definition of what constitutes the diversity of life.[21] Biodiversity certainly must include accounting for diversity of genotypes and phenotypes, the number and disparity of species, as well as their mutual interdependences. Biodiversity also includes consideration for the needs and effects of past and future species. Since species must be thought of always as integrated with habitats, the conceptualization of biodiversity must comprehend how biological multiplicity depends upon and can create diverse ecologies. As with definitions of species, the concept of biodiversity is open-ended, vague, often conflicting, and much politicized. Also, like species, biodiversity can include both natural and artifactual situations, not just because many ecosystems either involve human intervention or overlap with human developments and thus take on technological aspects. Many species can contribute to creating the ecological conditions for their own sustenance, thus "artificially" changing the system of nature they find themselves in. All these factors contribute to the need to continually be open to revising and challenging any theorization of biodiversity, rather than asserting a systemic categorization or metaphysical principle of life. As with thinking species, to think biodiversity one must consider how such thought could be revisable and redefined as the conditions of biodiversity change.

Although "biodiversity" is a recent term, the concept refers back to earlier notions in biological theory. Darwin famously finished *On the Origin of Species* with a description of what he called an "entangled bank":

> It is interesting to contemplate an entangled bank, clothed with many plants of many kinds, with birds singing on the bushes, with various insects flitting about, and with worms crawling through the damp earth, and to reflect that these elaborately constructed forms, so different from each other, and dependent on each other in so complex a manner, have all been produced by laws acting around us. These laws, taken in the largest sense, being Growth with Reproduction; Inheritance which is almost implied by reproduction; Variability from the indirect and direct action of the external conditions of life, and from use and disuse; a Ratio of Increase so high as to lead to a Struggle for Life, and as a consequence to Natural Selection, entailing Divergence of Character and the Extinction of less-improved forms.[22]

Understanding the entangled bank requires an entangled method—just as definitions of biodiversity require a diversity and plurality of methods and modes of knowing. Biodiversity is the diversity of biological ways of being, which are formed in biopolitics but are not reducible to biopolitics. Yet entangled and biodiverse species are not endlessly malleable—every species emerges from the bottleneck of previous species subject to laws of variability and selection and must relate to other species in complicated and shifting forms of dependence and independence. There is not endless or indefinite plasticity, but each change in form has costs and advantages.

Like the political concepts of "the people" and the commons, biodiversity is never filled but always being made. There is no guarantee that biodiversity will entail self-sustaining life, or even that care for biodiversity will set the parameters for a proper, normative standard of ethical engagement across all life. What biodiversity does offer is a platform by which multiple flourishings and multiple capabilities of living beings can be supported and extended. Since biodiversity cannot be defined permanently in advance or consolidated by any abstract structural or transhistorical principles, it has to be defined by its own event, by its own contestation. However, if there is only continual diversification, or difference, or even plasticity, as if these were preprogrammed for success, then biodiversity as a carefully entangled bank is meaningless, since these supposed properties of life would continue regardless of any fluctuations in apparent biodiversity. If there is only extinction, or if extinction is claimed to be the only ultimate biological reality that matters, then biodiversity or bioplasticity is again meaningless. However, if there is no extinction, because life is said to be made of intrinsically vitalist matter, here again biodiversity or bioplasticity is meaningless, since such vitality would be deemed independent of the constraints of existing forms and ecologies.

What biodiversity indexes, to me, is the "changing difference" that Malabou points to. The Anthropocene is precisely about trying to understand the shifting fates of species difference, plasticity, and finitude as entangled, opportunistic, and prone to failure. To think species, one has to think deconstruction and essence at the same time, not a natural essence, but an "empty essence" (to use Malabou's phrase) that is singular, formal, plastic, and emerges in relation to other singular and plastic species. Inquiring into the making and unmaking of species parallels an inquiry into considering a world without deconstruction even as we continue to use the tools of deconstruction. Certainly, we can imagine a world without deconstruction—and also a world without plasticity insofar as such a world is marked by the radical loss of forms. Indeed this is our world today—a plastic history that is both the history of changing forms and the loss of forms. Plasticity and deconstruction can help us understand both the proliferation of forms and the undoing of forms, but they both are embedded in a provisional plurality of ways of living that have no guarantees and can be radically changed either by ongoing plasticity or the erasure of forms. After deconstruction, form and essence regain importance as Malabou points out,[23] yet so does concern for the elimination of form, and for how the limitations of form both enable and constrain plasticity.

NOTES

1. Catherine Malabou, *Changing Difference: The Feminine and the Question of Philosophy*, trans. Carolyn Shread (Cambridge: Polity, 2011), 63.
2. Ibid., 87–88.
3. Jacques Derrida and Elizabeth Roudinesco, *For What Tomorrow . . .: A Dialogue*, trans. Jeff Fort (Palo Alto, CA: Stanford University Press, 2004), 68. Derrida insists that all consumption is at least symbolically carnivorous as it involves ingesting the other. But there is a significant portion of life, including single-celled organisms as well as most plants, which exist not by ingestion but by incorporation as transcorporation, or conversion of nonliving matter into another form. All feeding is conversion, a material exchange and transformation between internal and external bodies, but this is not necessarily carnivorism. The more general issue concerns how Derrida made his own categorical declarations at which point he stopped short of extending deconstruction or trying out new methods of thought.
4. Jacques Derrida, "The Deconstruction of Actuality: An Interview with Jacques Derrida," *Radical Philosophy* 68 (Autumn 1994): 38.

5. Jacques Derrida, *The Animal that Therefore I Am*, trans. David Wills (New York: Fordham University Press, 2008), 38.

6. Jacques Derrida, *The Beast and the Sovereign: Volume I*, trans. Geoffrey Bennington (Chicago: Chicago University Press, 2009), 15–16.

7. Derrida, *The Animal that Therefore I Am*, 30.

8. Charles Darwin, *Journal of Researches into the Natural History and Geology of the Countries Visited During the Voyage of H. M. S. Beagle Round the World*, 2nd edition (London: John Murray, 1845), 174.

9. Charles Darwin, *The Origin of Species* (Oxford: Oxford University Press, 1996), 45.

10. Noam Lahav, *Biogenesis: Theories of Life's Origin* (Cambridge, MA: Harvard University Press, 1999).

11. Michael Marder, *Plant-Thinking: A Philosophy of Vegetal Life* (New York: Columbia University Press, 2013), 7.

12. Gilles Deleuze, *Pure Immanence: Essays on A Life*, trans. Anne Boyman (New York: Zone Books, 2001), 27.

13. Ray Brassier, *Nihil Unbound: Enlightenment and Extinction* (New York: Palgrave Macmillan, 2007), 234.

14. Catherine Malabou, "Whither Materialism? Althusser/Darwin," in *Plastic Materialities: Politics, Legality, and Metamorphosis in the Work of Catherine Malabou*, eds. Brenna Bhandar and Jonathan Goldberg-Hiller (Durham, NC: Duke University Press, 2015), 47–60.

15. Darwin, *Origin of Species*, 67.

16. Malabou, "Whither Materialism?," 50.

17. Stephen Jay Gould, *Wonderful Life: The Burgess Shale and the Nature of History* (New York: Norton, 1989), 47.

18. Catherine Malabou, "Will Sovereignty Ever Be Deconstructed?," in *Plastic Materialities*, 38.

19. Elizabeth Kolbert, *The Sixth Extinction: An Unnatural History* (New York: Henry Holt, 2014), 130.

20. Malabou, *Changing Difference*, 65–66.

21. For philosophical discussions of the meaning of biodiversity, see James MacLaurin and Kim Sterelny, *What Is Biodiversity?* (Chicago: University of Chicago Press, 2008); Ronald L. Sandler, *The Ethics of Species: An Introduction* (Cambridge: Cambridge University Press, 2012).

22. Darwin, *Origin of Species*, 395–96.

23. Malabou, *Changing Difference*, 98.

BIBLIOGRAPHY

Darwin, Charles. *Journal of Researches into the Natural History and Geology of the Countries Visited During the Voyage of H. M. S. Beagle Round the World*. 2nd ed. London: John Murray, 1845.

———. *The Origin of Species*. Oxford: Oxford University Press, 1996.

Deleuze, Gilles. *Pure Immanence: Essays on a Life*. Translated by Anne Boyman. New York: Zone Books, 2001.
Derrida, Jacques. *The Animal that Therefore I Am*. Translated by David Wills. New York: Fordham University Press, 2008.
———. *The Beast and the Sovereign: Volume I*. Translated by Geoffrey Bennington. Chicago: Chicago University Press, 2009.
———. "The Deconstruction of Actuality: An Interview with Jacques Derrida." *Radical Philosophy* 68 (Autumn 1994).
Derrida, Jacques, and Elizabeth Roudinesco. *For What Tomorrow . . .: A Dialogue*. Translated by Jeff Fort. Palo Alto, CA: Stanford University Press, 2004.
Gould, Stephen Jay. *Wonderful Life: The Burgess Shale and the Nature of History*. New York: Norton, 1989.
Kolbert, Elizabeth. *The Sixth Extinction: An Unnatural History*. New York: Henry Holt, 2014.
Lahav, Noam. *Biogenesis: Theories of Life's Origin*. Cambridge, MA: Harvard University Press, 1999.
MacLaurin, James, and Kim Sterelny. *What Is Biodiversity?* Chicago: University of Chicago Press, 2008.
Malabou, Catherine. *Changing Difference: The Feminine and the Question of Philosophy*. Translated by Carolyn Shread. Cambridge: Polity, 2011.
———. "Whither Materialism? Althusser/Darwin." In *Plastic Materialities: Politics, Legality, and Metamorphosis in the Work of Catherine Malabou*, edited by Brenna Bhandar and Jonathan Goldberg-Hiller, 47–60. Durham, NC: Duke University Press, 2015.
———. "Will Sovereignty Ever Be Deconstructed?" In *Plastic Materialities*, 35–46.
Marder, Michael. *Plant-Thinking: A Philosophy of Vegetal Life*. New York: Columbia University Press, 2013.
Ray, Brassier. *Nihil Unbound: Enlightenment and Extinction*. New York: Palgrave Macmillan, 2007.
Sandler, Ronald L. *The Ethics of Species: An Introduction*. Cambridge: Cambridge University Press, 2012.

Chapter 9

The Plasticity of Empathy

A Materialist, Postphenomenological Critique of Einfühlung *in Aesthetics, Phenomenology, and Contemporary Neuroscience*

Andrew Bevan

This chapter describes a historical development of the concept *Einfühlung* (translated into English with the neologism "empathy" in 1909) and its invocation in aesthetics, phenomenology, and contemporary neuroscience to explain intersubjectivity. Both *Einfühlung* and empathy are composed of the prefix *ein-* or *em-* to typically denote a unidirectional movement of feeling (*fühlung* or pathos) "from" an immutable and universal human unaffected by its projection "into" a purely receptive exteriority to understand this "to" appended secondarily to the privileged "from." This fails to acknowledge an originary simultaneity of "from" and "to" in which both become radically mutable to produce the difference between "from" and "to." Applying Malabou's concept of plasticity to empathy better acknowledges this mutability and resulting radical formative formations of the human and modes of intersubjective understanding without universality, norm, or limit.

ROOTS OF *EINFÜHLUNG*

The roots of *Einfühlung* in German Romantic philosophy attest to an early privileging of human form and unidirectionality. Herder used *einfühlen* to describe the ascription of human feelings to natural phenomena resembling human form, a mystical union of human with nature, without explaining how such feelings or form arise or develop. Herder also used the concept for textual interpretation where, to fully understand texts, readers were exhorted to "go into the age, into the clime, the whole history, feel yourself into [*einfühlen*] everything."[1] But, to respect the alterity of the "into," Herder stressed the difficulty of grasping a contemporary human let alone those of another age: between experiencing an other as they experience themselves there lies "a gulf."[2]

AESTHETICS: VISCHER AND *EINFÜHLUNG*

However, it was in German aesthetics that *Einfühlung* would emerge as a concept with Robert Vischer the first to use it as a substantive in his dissertation On the Optical Sense of Form (1873).[3] Vischer's father, Friedrich, had argued forms devoid of emotional life have their emotional content supplied by us, a process he called symbolism of form [*Formsymbolik*], a "rapt, inward sense [*Ineinsfühlen*] of the unity of image and content."[4] Robert now sought to elaborate this *Ineinsfühlen*.

Vischer developed Köstlin's theory of "association of ideas"—where just as one external form reminds us of another so we find resemblances between external things and our internal states due to our capacity for reflexivity—by adding another process, "a direct merger [*Verschmelzung*] of the imagination with objective form."[5] Vischer found this idea in Scherner's book on dreams (*Das Leben des Traumes*) that posited sensations arising from stimulated bodily parts while asleep were represented in dreams by analogous objects. Both these processes, however, are separate to another "psychically necessary" symbolism of form intrinsic to human imagination in which "we maintain our freedom to perceive the symbolic process as nothing more than an analogy."[6]

Vischer begins with sensation [*Empfindung*], "the most primitive impulse of life" out of which "evolve the more distinct acts of the imagination, volition, and cognition."[7] Sensation arises because "every mental act is brought about and is at the same time [*zugleich*] reflected in certain vibrations and who knows what neural modifications, in such a way that the latter represents its image, that is to say, they produce a symbolic picture inside the organism."[8] The external object "accords with the related vibrations, stimulates and confirms them, and with that the emotional state reflects itself in them."[9] Vischer marks a difference between the sensory or immediate [indicated by the prefix *Zu-*] and the motoric or responsive [*Nach-*]. Applied to sensation, the immediate becomes "the passive process of a sensory or pure nerve function," the responsive "the active process of a motor-nerve function—a muscular movement"[10] with responsive sensation [*Nachempfindung*] following immediate sensation [*Zuempfindung*].

Sensations as vibrations occur through "similarity" [*Aehnlichkeit*], a harmony between subject and object that arises "because the object has a harmonious form and formal effect corresponding to subjective harmony."[11] Immediate visual sensations depend on "the similarity or dissimilarity of the object, first with regard to the structure of the eye and second with regard to the structure of the whole body"[12] because visual stimuli affect the whole body. Pleasant sensations occur with objects more easily aligned with our sensory faculties: "the horizontal line is pleasing because our eyes are positioned horizontally"[13] whereas an oblique line, requiring uncomfortable movements, produces unpleasant sensations. Vischer's similarity thus posits physiological laws, "subjective laws of the normal human body,"[14] that also subsume those of symmetry and proportion, where law is "the only true repose" and "the stable form or ideal frame within which motion occurs."[15]

Vischer now turns to the imagination [*Bildvorstellung*], "the act by which we mentally simulate something that previously existed as a vague content of our sensation."[16] I can imagine objects [*Objektsvorstellung*] or my self [*Selbstvorstellung*] but imagining my self only becomes conscious "when it relates itself to an object or idea of the object."[17] Scherner had shown how daily experience provided the objects which stimuli merge with in dreams—a head hanging out of bed might be symbolized by "a dangerously overhanging oriel of a house"[18]—and Vischer now extends this merging to waking imagination. Imagination is thus a hybrid of sensation and object, "a fluid medium in which contradictions of the world—repose and motion, self and nonself—merge into a mysterious whole."[19]

While immediate and responsive sensations can remain external to an object they can, with the aid of the imagination, also move *into* an object, *Einempfindung*, to deepen sensation. But sensation only intensifies into feeling [*Gefühl*] by perceiving "a spiritual value or vital force" in the appearance; feeling requires another human as self-feeling is dull and sterile so "strives on its own accord to reach out beyond itself and yearns for a reciprocal feeling elsewhere" because of a "natural love for my species" that enables my projection in which I ostensibly "feel not only myself but at the same time the feeling of another being."[20] Imagination with sensation enabled comparison with objects, imagination with feeling enables comparison with a spiritual core. Instead of objects sensed as hindrances or conveniences, "there now appears a living individual or a community of individuals that either sympathizes with and supports us . . . or unfeelingly and spitefully works against us"[21] and the relation between subject and object deepens into one between subject and subject.

Vischer therefore seems to be arguing inanimate objects play no role in feeling, that we only *feel* an appearance when it "supplies the language of feeling from its own resources": "What are all those forms to me through which the red blood of life does not flow? I do not measure my heart with the same yardstick as I do a lump of stone."[22] But, remarkably, Vischer argues it is because of this absence of life that we project it into inanimate objects: "our feeling rises up and takes the intellect at its word: yes, we miss red-blooded life, and precisely because we miss it, we imagine the dead form as living."[23]

Immediate sensations now intensify into immediate feelings [*Zufühlung*] which "affect the whole mood [*Stimmung*] of the person" where we are "charmed or repelled" without knowing why—warm objects best exemplify being charmed as we identify them with the warmth of human feeling.[24] Responsive sensations intensify into responsive feeling [*Nachfühlung*] through acts of self-movement: when I look at a winding road, my thoughts track its movement: "I seek and find, ascend triumphantly and fall to destruction," this movement thus emulating "human impulses and passions."[25]

As with sensation, feeling can also be projected *into* the inner structure of the object, *Einfühlung*: "I can think my way into it, mediate its size with my own, stretch and expand, bend and confine myself to it."[26] It can be motionless, a physiognomic or moodful *Einfühlung* of involuntary habit where, when the object is small, I experience a static contractive feeling [*Zusammenfühlung*], when large, a static expansive feeling [*Ausfühlung*]. Or it can be moving, a "mimicking, acting, or affectful *Einfühlung*" that projects motion into both moving and static objects by treating the static as "apparently moved," one we imagine has just moved or is about to move, like a cliff that "appears to stand at attention and squarely face us . . . its projecting angle seems to lunge out as if affected by a passion (impatience, curiosity, anger)."[27] There is thus also a moving contractive and expansive feeling: contractive "synonymous with a weakening or renunciation of the self," expansive "with a strengthening and liberation."[28] While concerned with how we supply inanimate forms with emotional content, Vischer admits *Einfühlung* also projects into organic nature to animate a plant or anthropomorphize an animal and only with other human beings does *Einfühlung* act as "a doubling of self."[29]

Vischer clarifies the difference between *zu-*, *nach-*, and *ein-* prefixes. Immediate feeling, *Zufühlung*, functions most directly, "a direct intellectualization of sensory stimuli"[30] that causes a responsive feeling, *Nachfühlung*, in real or imagined motor activity that is still simple and direct when compared to *Einfühlung*. Both *Zufühlung* and *Einfühlung* leave the self in a sense solitary because "I attend solely to the object" but whereas both zu- and nach- move "from the outside (the object's form) to the inside (the object's center)," *Einfühlung* moves from inside to outside.[31] *Einfühlung* "takes on a life of its own" and there occurs a "central projection, exchange and return" where *Einfühlung* "looks at the second self as it sits reshaped in the object and intuitively takes it back to itself, yet without discerning it clearly or knowing why."[32] These symbolizations work first with each other and then with association of ideas, an "absolute interlacing and togetherness" that produces "a true aesthetic appreciation of form."[33]

The "remarkable merger [*Verschmelzung*] of subject and object" in feeling-imagination [*Gefühlsvorstellung*] occurs because the nature [*Wesen*] of Vischer's feeling is "a pantheistic urge for union with the world, which cannot be limited to our more easily understood kinship with the human species but must, consciously or unconsciously, be directed toward the universe."[34] Vischer adds, "I feel in order to enjoy the universal in me or in the world, and this imagined perfection of the universe easily bears me beyond the faltering and difficult fate of the individual."[35] *Einfühlung* thus arises from an intolerable otherness that motivates mergers with objects but which loses the self, a simultaneous alternating attachment to self and other that generates successive mergers and separations, an "alternate masking and ummasking of the human form"[36] which has a tendency of "confusing our own feelings with that of nature"[37] and confusing "our own stimulation with the thing that produces the stimulus."[38]

To overcome this confusion, feeling must intensify into *Gemüth*, a word difficult to translate, formed from *ge*-signifying a unity and *Muth* for mind or state but here problematically translated as emotion. It does this "the more we become aware of universal coherence [*Weltzusammenhang*]."[39] Cosensation or cofeeling [*Mitempfindung, Mitgefühl*] we might have with an individual deepens to a more profound kinship with the entire human race, uniting individual with universal. Projection of feeling into nature now occurs subtly and unconsciously until the projected body is forgotten in favor of the soul [*Seele*] projected with it and "anything and everything" becomes spirit [*Geist*].[40] Continuing his pantheism, Vischer argues that, as a result, man comes to believe either in a power residing in things or in god(s) directing these things. But, although the imagination will always "secretly suggest the organic forms and movements of its own body,"[41] through a strict emphasis on the purely aesthetic symbolism of form we can learn to separate the objective form of the external phenomena from that which the imagination has projected into it. Through this separation, Vischer adds, the ideality of imagination emerges.

Vischer concludes by relating this to art. Artistic transformation occurs when I "replicate for my fellow human beings what goes on inside me."[42] Art becomes an imitation not of the object but of the artist's empathic projections into the object. The artist transcribes these movements into a more permanent material, returning them to their original state of unity and simultaneity that continues to evoke these movements in the viewer by this same symbolism of form, like the "stormy character and rushing energy of Ruben's technique."[43] The artist transposes the "intuitive and rational human norm to the object" that describes "the perfect human being" and "harmonizes the phenomenon so that in his hand it grows into a complete expression of the felt human harmony and divinity of the universe."[44] Aesthetic creation and appreciation thus neither begins nor ends with the act of creation but has its origin and continuation in a dynamic, emulating process involving self (imagination with sensation, feeling and *Gemüth*) and object, simultaneously a practice and an intensification of sensation toward universal coherence and perfection.

But Vischer's distinction between an aesthetic symbolism of form and an intrinsic one from which to view any symbolism as mere analogy, a position seemingly free of such analogizing, nevertheless still requires both to give the difference between the two and the ability to separate external phenomena from the body projected into it, a separation that nevertheless retained the projection of soul. Absolute separation thus seems impossible, or at least intolerable, and some merger of "from" and "to" persists across separations. Thus, any "from" that would be unaffected by the "to" with which it is merged, capable of being understood without this "to," or act as place of a "return" unaffected by projection and exchange, any immutable realm of pure ipseity is problematized. The absence of such undermines any real distinction between an aesthetic and intrinsic symbolism of form, both require otherness.

If there is no "from" without "to," and given Vischer's concept of similarity suggested differences in likelihoods of things acting as "to," this "to" can no longer be considered purely passive, purely receptive to projections of the "from." Unless posited as static, changes in this "to" would affect mergers of "from" and "to" as well as likelihoods of objects acting as "to" in subsequent mergers. An actively changing "to" affects objective harmony to better correlate with and reinforce the "static" laws of subjective harmony. The "from" would thus be most active when merging with most dissimilar objects, mergers that, without any immutable realm, could affect subjective harmony. A more active "from" producing difference, more passive sameness.

Finally, this activity and passivity on both sides undermines the implicit unidirectionality of *Einfühlung*. The "from" and "to" would therefore be better conceived as an originary simultaneity, dyadic at minimum, where any directionality does not precede but is instead produced with separations that posit a "from" and "to" as differences in activity and passivity, like the merger of imagination and object that separates as a projection of active imagination into passive external phenomena, separations that nevertheless remain a merger, a different simultaneity. And let us note that any simultaneity remains simultaneous with other simultaneities; as Vischer states, "the brain itself functions on many levels."[45]

With these qualifications, Vischer could be read not merely as an account of how the inanimate is supplied with emotional content but also as the means by which emotional life is understood and develops, an intensification of originary sensation into feeling and *Gemüth* that differentiates the ideality of imagination and the reality of external phenomena, self and other. Any empathizing, any "projection" of feeling "into" another human, participates in the same process: an alternating simultaneity of intolerance of otherness and attachment to self produces oscillations in dyadic confusions of self with other, separating as different simultaneities of self now with and other without that which was projected.

Such a reading would reveal significant differences in intensifications and differentiations of the "from," differing with each "to" (animate/inanimate, organic/nonorganic, human/nonhuman, static/moving, real/imaginary, absent/present) to produce different sensation(s) and feeling(s) (immediate/responsive, contractive/expansive, static/moving, furthering/disturbing) and radically diverse forms and understandings of "from" and "to," self and other. Differences that modify relations to the subsequently separated through "passions of desire or recoil"[46] and which, as reflexivity offers aspects of self as possible mergers, self as both "from" and "to," also applies to the relation to self. The absence of an immutable "from" or "to," both active and passive, mergers contingent on environment and experience, a process without end, lacking any absolute separation, also undermines Vischer's originary or ultimate universal coherence or perfectibility.

LIPPS: *EINFÜHLUNG* AND INTERSUBJECTIVITY

It was Theodor Lipps who first invoked *Einfühlung* to explicitly explain intersubjective understanding. In *Das Wissen von Fremden Ichen* (1907),[47] Lipps tackled the problem both of how I understand myself as a self, a specific "I" among other "I"s, and how I understand an other as other. Lipps begins by rejecting the "absurdity" of the prevailing account, inference by analogy [*Analogieschluss*], that argued the behavior of others is understood by comparing it to our own—my happiness is accompanied by bodily changes so when I see the same changes in others I infer, on the basis of their similarity to me, that they too are happy. Lipps believed this only described how another's happiness reminds me of my own, it does not explain the attribution of happiness to the other, attribution would require that

> I am not meant to think again my own grief or rage, but I am supposed to think something absolutely different, that is, instead of myself and my own grief or my own rage, the grief or rage of another person. I am supposed to swap myself, the absolute subject, against something that is for me object and only object.[48]

Lipps developed *Einfühlung* to explain this attribution characterizing it as an "original and irreducible" instinct comprised of two factors: an instinct for expression or "manifestation of life" and an instinct for mimicry.[49] The instinct for expression means that when I experience an emotion or feeling, it is externally manifested in the activity of my body, particularly its facial expressions. Now, when this expression is perceived by another, the instinct of imitation is invoked so that, not only do they reimagine the feeling, they also relive it: "when I see a gesture, the tendency to experience the affect, out of which this gesture naturally evolves, exists within me. . . . I experience thus the inner state, which I can see another express, in me."[50] Lipps then adds an extra, final stage to create intersubjective understanding: I *think* my affect *into* the gesture of the other (*Hineindenken*).[51]

PHENOMENOLOGY AND *EINFÜHLUNG*

Critique of Lipps

Next is the development of *Einfühlung* by the phenomenologists Scheler, Stein, and Husserl who rejected Lipps's psychologism, criticized its assumptions and sought a radical new grounding for intersubjective understanding. Lipps, they argued, assumed I already understand my own feelings before those of the other and express them in a similar manner, that Lipps's "natural" instinct for expression ignores dissimulation or suppression of feeling and that full understanding of the other's expressions is undermined by my inability to see my body the way the other sees it. Scheler gave examples of understanding without experiencing the feeling: "a person who has never undergone mortal terror can still understand and envisage it, just as he can also share in it"[52] and I can understand a dog's happiness by its wagging tail despite having no tail to wag.[53]

In *On the Problem of Empathy* (1917) Stein argued Lipps had described not *Einfühlung* but a form of emotional "contagion" or "transference."[54] Instead, Stein focused on the seemingly paradoxical experience of *Einfühlung*, whereby "this other subject is primordial although I do not experience it as primordial. In my nonprimordial experience I feel, as it were, led by a primordial one not experienced by me but still there, manifesting itself in my nonprimordial experience."[55] Stein, echoing Vischer's alternations in dyadic confusions of self and other, adds that when I empathize into another's place, I "transition from my viewpoint to the other's" but "[t]he same world is not merely presented now in one way and then in another, but in both ways at the same time."[56]

Husserl and *Einfühlung*

Husserl next developed (although later abandoned) *Einfühlung* in *Cartesian Meditations*, again as a means of explaining intersubjectivity, describing *Einfühlung* as having received its "true sense" and "the true method for its solution."[57]

Husserl begins by considering how an "alter ego" appears as an intentional object and the two ways of experiencing them: either as objects "'*in' the world*," things among other things, or as "*subjects for this world*" experiencing the world as I do. Husserl uses a distinction in German between *Körper* for the former, for any body, and *Leib* for the body of an animated organism. Other animate organisms are experienced as "*governing psychically* in their respective natural organisms" and I therefore experience the world as "other than mine alone, as an intersubjective world, actually there for everyone."[58] Husserl must now develop "a *transcendental theory of experiencing someone else*, a transcendental theory of so-called 'empathy.'"[59]

First "*a peculiar kind of epoché*" is enacted to "*disregard all constitutional effects of intentionality relating immediately or mediately to other subjectivity*," limiting oneself to a "sphere of peculiar ownness" constituted within.[60] This yields a "founding stratum" in which nothing other exists, where my own *Leib* is "uniquely singled out" as the only *Körper* that is a *Leib* because it is "reflexively related to itself," a capacity for autoaffection that enables me to "ascribe fields of sensation" in which I "'rule and govern' immediately."[61] Husserl emphasizes all the things my *Leib* can do: I can perceive with my hands, my eyes, I can push, thrust and therefore act somatically. This constitutes my transcendental Ego and gives my *Leib*, my psyche [*Seele*] and their psychophysical unity through which "my personal Ego operates in this animate organism and, "by means of" it, in the "external world."[62] Husserl describes the analyses in this level as "transcendental aesthetics."[63]

The next level is the moment something other confronts us in this reduced sphere of ownness. The first other is "the other Ego" appearing as an "immanent transcendency" and Husserl must now clarify "experiencing 'someone else' in the sense in which the other has not yet attained the sense 'man.'"[64] Although the only animate organism in this sphere is my own, the other is nevertheless apprehended as animate because of *an apperceptive transfer from my animate organism*[65] based on its similarity to me. This is not inference by analogy because apperception "is not inference, not a thinking act."[66] There now occurs a continuous and reciprocal "pairing" of my ego with the alter ego with an "overlaying of each with the objective sense of the other" such that "a mutual transfer of sense" occurs.[67] Contrastive pairings add somatic content to my transcendental ego through "the understanding of the other's organism and specifically organismal conduct: the understanding of the members as hands groping or functioning in pushing, as feet functioning in walking, as eyes functioning in seeing, and so forth."[68]

Husserl's next level is "an 'empathizing' belonging to the 'higher psychic sphere'" based again on "my own conduct under similar circumstances" such as "the outward conduct of someone who is angry or cheerful, which I easily understand from my own conduct under similar circumstances."[69]

However, the other only *appresents* their psychic life like a physical thing appresents a hidden, reverse side but whereas the object's appresentation is primordially fulfillable by rotation to view this other side, the other's psychic life requires continual verification with them because its appresentations are not primordially fulfillable, I can perform no rotation to view their hidden psychic life; if I could, "it would be merely a moment of my own essence, and ultimately he himself and I myself would be the same."[70] I must therefore continually verify my assumptions about their psychic life with them. Higher and diverse psychic occurrences thus become understood through "synthetic interconnexions" and *Einfühlung* is now explicitly simultaneous, reciprocal, and continuous:

> every successful understanding of what occurs in others has the effect of opening up new associations and new possibilities of understanding; and conversely, since every pairing association is reciprocal, every such understanding uncovers my own psychic life in its similarity and difference and, by bringing new features into prominence, makes it fruitful for new associations.[71]

But why was the first experience of something other that of experiencing "*someone* else"? If this other "has not yet attained the sense 'man'" and I have disregarded all constitutional effects from other subjectivities why is it not experiencing *something* else, something that may or may not attain the sense "man"?[72] Has not Vischer demonstrated the diversity in possible mergers, of myself and any*thing*, differences that would affect possibilities for pairing and empathizing and the subsequent somatic and psychic content added? That the first other is another ego arises because Husserl treats the *Körper* in the founding sphere of ownness as a "human" *Körper*, he does not truly disregard all constitutional effects for I still know the names and divisions of my body, can differentiate the senses and, perhaps most damning, retain language. Are these not constitutional effects relating to other subjectivities? If I could truly disregard such effects would that not render impossible recognition of other subjectivities as similar?

This for Derrida, one of Husserl's keenest critics, marks the anthropological limits of phenomenology: "for all its ambition and originality, phenomenological reduction here does not suspend the Ego's human appurtenance or determination."[73] One might reply one cannot deny the body's reflexivity but Derrida counters reflexivity also requires a "detour by way of the foreign outside,"[74] a detour through this (not yet human) *Körper* that implies a "spacing" between me and myself, which enables me to say *my Körper*, a unity of my psychic and physical which makes it my *Leib*. Derrida therefore asks, "Shouldn't a certain introjective empathy, a certain 'intersubjectivity,' already have introduced an other . . . to give rise to an experience of the body proper allowing one to say, 'it is I,' 'this is my body?' "[75] Derrida thus characterizes the body's reflexivity not as pure autoaffection but "autoheteroaffection" where something simultaneously other (hetero) accompanies any ascription of sensation, something passively formed with active autoaffection. Derrida adds, "the inanimate, 'material nature,' as well as death, the nonliving, the nonpsychical in general, language, rhetoric, technics and so forth"[76] should thus be reintroduced into this phenomenological sphere of ownness.

Husserl's three levels resemble those of Vischer's intensification of sensation into feeling and *Gemüth* but in question again is the possibility of a realm free of otherness. Husserl explicitly argues for such when he states of the sphere of ownness, "I obviously cannot have the 'alien' or 'other' as experience, and therefore cannot have the sense 'Objective world' as an experiential sense, without having this stratum in actual experience; whereas the reverse is not the case."[77] That is, I must have my sphere of peculiar ownness to experience something other whereas I do not need something other to experience my own sphere. Rejecting this realm (as with Vischer) and thus any possibility of pure phenomenological reduction, Husserl's careful analyses of the simultaneous, reciprocal, and continuous nature of empathy in the development of self and other could be retained, together with Derrida's "introjective empathy," to better recognize a radical alterity and mutability in both. Rejecting any anthropological limit, the first other could be any*thing*, not necessarily human (however understood), and the difference between human and nonhuman may not precede but is instead produced, although not guaranteed, by pairings with these *things*. After Vischer's diverse descriptions of mergers with the inanimate, what differences in foundations and subsequent possibilities of pairings, empathizing and intersubjective understanding could arise?

HETERODOX FORMINGS

An example of one such case is perhaps provided by the psychoanalytic case study of the "crane-child" by François Péraldi.[78] It is the story of Michel, the child of a mother raped by a drunkard at sixteen, "a poor woman, completely disorientated in life and in a world she has never understood" who "would sit in silence, waiting for some words to come out of the tiny mouth and, when she would bend over the body of her child, she presented him only with the marmorean mirror of her face—a mirror in which Michel could recognize himself as if he were of stone."[79] Through the window of Michel's cot, he could see a building site where cranes worked, "he could hear them talking to him, for they did not wait for him to speak first, they just were 'talking,' mixing repetitive gnashings, gratings, grindings with the orders shouted by invisible men."[80] As a result, in the institution where Michel now resides,

Péraldi states Michel "talks about only one thing: cranes. . . . He imitates, on all sorts of semiotic levels (voice, gesture, noises as well as spoken language) only one thing: cranes."[81]

Péraldi relates the difficulty of communicating with Michel. First, he tried a rigid oedipal psychoanalytic framework (an attempt that was met by Michel destroying several buildings) but later questioned this approach: "instead of trying to elaborate better techniques to make him enter into the realm of Oedipus, we suddenly realized that we had not been listening to Michel with the not-knowing ear, that we had refused to listen to the *singularity* of a desire that had nothing to do with ours."[82] Thus, they rejected abstract preexisting frameworks to better find a common ground with patients, to analyze their own desire and "to listen to it as it responded to the desire of the psychotic."[83] This led to changes in both analysts and patients who developed "not toward normality, if such a name ever had any other meaning than conformity to the law, but rather toward what I call a complexification of their relationship to their *Umwelt*, to their environment, to the others."[84] This complexification led to "a multiplication and diversification of the semiotic levels and of the semiotic procedures that are usually either refused or prohibited in normal communication."[85]

Can this "multiplication and diversification" be compared with Husserl's pairing associations and spreading "synthetic interconnexions" which reciprocally uncovers life, opening possibilities for new associations? If so, Péraldi's questioning of modes of communication and rigid frameworks unilaterally established by a normative "from" reveals intersubjective communication cannot be assumed but must tirelessly be sought, each time for the first time, acknowledging both sides may complexify and diversify as well as the possibility one may not be achieved.

For we are not born into a world purely receptive to our formative acts, nor are we purely unreceptive to the world, a world exceeding our birth and death, a world already racialized, gendered, abled that affects possibilities of intersubjective understanding. Fanon provides another example of this differently accommodating world. Fanon initially accepted the concept of bodily schemas as "implicit knowledge" arising from habituated relations within the world, "a slow composition of my self as a body in the middle of a spatial and temporal world . . . a real dialectic between my body and the world"[86] arising through "residual sensations and perceptions primarily of a tactile, vestibular, kinesthetic, and visual character."[87] But, in the experience of encountering a child in the street who shouts, "Mama, see the Negro! I'm frightened!," Fanon writes, "The corporeal schema crumbled, its place taken by a racial epidermal schema."[88] Fanon "wanted to come lithe and young into a world that was ours and to help to build it together"[89] but instead finds himself, "an object in the midst of other objects"[90] an object determined by a world that exceeds him. Beyond any corporeal schema, Fanon finds himself slave to a "historico-racial schema" formed "by the other, the white man, who had woven me out of a thousand details, anecdotes, stories."[91] Any universal, equal human form acting as site of ethical understanding is hopelessly naive; no such neutral ground exists on which all bodies align, bodies instead formed and forming in their detour through an outside that takes in centuries of racist history, gender prejudice, and dominant beliefs about the "normal."

PLASTICITY

All these criticisms can be incorporated succinctly with plasticity. Malabou defines the first two senses of plasticity: "it means at once the capacity to receive form (clay is called "plastic," for example) and the capacity to give form (as in the plastic arts or in plastic surgery)."[92] The plasticity of the human brain, though, is more complex; it is "something modifiable, 'formable,' and formative *at the same time*."[93] But plasticity also has a third meaning, "the capacity to annihilate the very form it is able to receive or create" as evidenced by "plastic explosives."[94] Thus, Malabou says, to talk of the plasticity of the human brain means "to see in it not only the creator and receiver of form but also an agency of disobedience to every constituted form, a refusal to submit to a model."[95] Plasticity can better conceptualize contingent encounters of "from" and "to": while many describe a human "from" actively forming its "to" and some hint at this "to" actively forming "from," plasticity adds that, in actively forming this "to," the human simultaneously passively forms itself. A human "from" or "to" is thus itself an originary simultaneity of "from" and "to" that meets an active/passive "to" or "from." Encounters between two humans therefore entail more complex, reciprocally affecting encounters with both sides capable of forming and being formed simultaneously which further undermines any simple claim to universal intersubjective understanding or originary or final coherence, any active *cohering* simultaneously passively cohering itself differently affecting possibilities of others cohering, any coherence also liable to annihilate its existing coherence.

To better recognize the most radical diversity in possible mergers and pairings of "from" and "to" the field of formative materiality should be flattened. This necessitates rejecting historic privilegings of the "from" over the "to": psyche over physis, self over other, human over nonhuman, animate over inanimate. But how to affirm a flat ontology of formativity while simultaneously acknowledging the hierarchization in formations observed with Péraldi or Fanon? Dominant formations form nonetheless, normativizing and sedimenting their proclaimed universality. How to produce a "split" or double ontology?

The lead here will be the simultaneity of receiving and giving form in plasticity which, together with its potential to explode form, ultimately reveals, as Malabou argues, "the schizological tendency of ontology [which] enables the dialecticizing of the simple, breaking its sculptural effect."[96] This means the possibility of change or destruction is already in the form that comes to presence. Any attempt to reside in a pure self, free of analogy or otherness, merely forms differently in this attempt at purity. This schizo-ontology applies to the inanimate and animate, immaterial and material, faculties of sensation, feeling or imagination, individual sensations and feelings and objects to affirm any merger or pairing, any simultaneity of activity and passivity *is* the mark of being this active and passive binding.

CONTEMPORARY DEBATE

However, the radical implications of plasticity are rarely acknowledged in contemporary discussions of empathy. Some neuroscientific attempts to explain empathy's role in intersubjectivity are accused of depending on a "basic emotions" view that assumes universal, innate emotions such as fear, anger, and disgust are manifested in distinct physiological expressions of authentic emotional experience localized in specific brain pathways. The experience of understanding emotional expression is often depicted as if unmediated by other cognitive states assuming, particularly in the contentious evidence for mirror neurons, an immediate, innate, and static mutual attunement of emotion between people.[97] Others, however, particularly in the humanities, reject this account in favor of emotional expression that is entirely culturally specific, mediated by its context but who are then accused by neuroscientists of ignoring its evidence for common pathways in the emotional brain.[98]

But this debate mostly repeats the debate between Lipps and Phenomenology a hundred years ago. It persists for three main reasons. First, few adequately consider how specific sensations, feelings, or emotions nor the faculties themselves of sensation, feeling, or emotion arise and change, all are accepted too often as given without clear definition or distinction, as "transcendental signifiers" that structure the debate. This despite the diversity in conceptions of sensation(s), feeling(s), or emotion(s) demonstrating their very plasticity. Second, implicit in most attempts to think *Einfühlung* and empathy has been the assumption of the possibility of pure autoaffection, a realm of pure ipseity from which to project into the world to understand it and into which one can always return, unaffected by this projection. Third, there is a resistance to abandoning the presupposition of an innate understanding of the human/nonhuman and animate/inanimate difference or innate orientation to "human" and thus to thinking the radical implications of its absence.

While there are undoubtedly common expressions of common emotions and common understandings of what those expressions express that together facilitate widespread intersubjective understanding, the alternative conceptualization presented here can also affirm the coexistence of heterodox emotions and expressions, radically different simultaneities of sensation(s), feeling(s), or emotion(s) that, with contingent encounters of heterodox and orthodox, reciprocally affect each other to further diversify and forestall universal intersubjective understanding. Attempts to invoke sensation(s), feeling(s), or emotion(s) as ground for this understanding merely participate in this same process of formative formation. Replacing autoaffection with autoheteroaffection, any human "from" of pure ipseity is rejected for a radically mutable simultaneity of "from" and "to" alternating in relative activity and passivity to produce the difference between "from" and "to" in contingent encounters with an other. A split ontology, simultaneously flat and hierarchized, better acknowledges this radical diversity in formations of the human without primary orientation to or understanding of the "human," its disjuncture explaining the difference in relative frequency and universality with which emotions expressed sediment and habituate in their continuous spreading of associations and appresentations, a flat formativity forming a communally experienced world more and less accommodating to the formations whose formativity in turn widens and narrows possibilities of intersubjective understanding. Instead of any assumed universal "we" let us instead posit only a universal struggle to achieve a "we," a struggle not toward some ultimate universal coherence but toward unknown plastic potentials.

To the debate between humanities and sciences, it should be clearer now that cultural formative formations do not meet a purely receptive, purely mutable human or world nor do innate formative formations meet a purely unreceptive, purely immutable world or human. With the paradigm shifts of plasticity and epigenetics any immutable human "from" or "to" is undermined but limits to plasticity and evidence for some determining genes mean this is not a pure mutability. The debate could then shift to the more productive question of how the coexistence of expressive emotions of greater and lesser evolutionary age, frequency, universality, orthodoxy, or intelligibility reciprocally affect each other and bring the humanities and the sciences in the "affective turn" into a more fruitful, one could say empathic simultaneity, to uncover radical associational possibilities. This alternative conceptualization also means that, rather than take sides in the hackneyed debate of essentialism versus antiessentialism, innate versus constructed, natural versus cultural, their simultaneity can instead be affirmed. For, ultimately, this is the lesson of plasticity and epigenetics: that construction is innate, antiessentialism is essential, and to be genetically determined means not to be genetically determined.

NOTES

1. Johann Gottfried Herder, *Herder: Philosophical Writings* (Cambridge: Cambridge University Press, 2002), 292.
2. Ibid.
3. Robert Vischer, "On the Optical Sense of Form," in *Empathy, Form and Space*, ed. H. F. Mallgrave and Eleftherios Ikonomou (Santa Monica, CA: The Getty Center for the History of Art and the Humanities, 1994).
4. Ibid., 90.
5. Ibid., 92.
6. Ibid., 90.
7. Ibid., 109.
8. Ibid.
9. Ibid., 90.
10. Ibid., 95.
11. Ibid.
12. Ibid., 97.
13. Ibid.
14. Ibid., 98.
15. Ibid., 95.
16. Ibid., 90.
17. Ibid., 100.
18. Ibid., 100.
19. Ibid., 102.
20. Ibid., 103.
21. Ibid.
22. Ibid.
23. Ibid., 104.
24. Ibid., 107.
25. Ibid.
26. Ibid., 104.
27. Ibid., 105.
28. Ibid., 105.
29. Ibid., 106.
30. Ibid., 108.
31. Ibid.
32. Ibid.
33. Ibid., 109.
34. Ibid.
35. Ibid., 112.
36. Ibid., 111.
37. Ibid., 107.
38. Ibid., 108.
39. Ibid., 109.
40. Ibid., 112.

41. Ibid.
42. Ibid., 114.
43. Ibid., 118.
44. Ibid., 115.
45. Ibid., 99.
46. Ibid.
47. Theodor Lipps, "Das Wissen von Fremden Ichen," in *Leitfaden der Psychologie* (W. Engelmann, 1906), translations my own.
48. Ibid., 708.
49. Ibid., 713.
50. Ibid., 719.
51. Ibid., 717.
52. Max Scheler, *The Nature of Sympathy* (Piscataway, NJ: Transaction Publishers, 1970), 47.
53. Ibid., 11.
54. Edith Stein, *On the Problem of Empathy* (Washington, DC: ICS Publications, 1989), 23.
55. Ibid., 11.
56. Ibid., 64.
57. Edmund Husserl, *Cartesian Meditations* (The Hague: Martinus Nijhoff, 1960), 147.
58. Ibid.
59. Ibid., 91–92.
60. Ibid., 93.
61. Ibid., 96–97.
62. Ibid.
63. Ibid., 146.
64. Ibid., 107– 108.
65. Ibid., 110.
66. Ibid., 111.
67. Ibid., 113.
68. Ibid., 119.
69. Ibid., 120.
70. Ibid., 109.
71. Ibid., 120.
72. Ibid., 108.
73. Jacques Derrida, *On Touching: Jean-Luc Nancy* (Palo Alto, CA: Stanford University Press, 2005), 164.
74. Ibid., 175.
75. Ibid., 176–77.
76. Ibid., 180.
77. Husserl, 96.
78. François Péraldi, "The Crane-Child," in *Psychoanalysis, Creativity and Literature: A French-American Inquiry*, ed. Alan Roland (New York: Columbia University Press, 1978).

79. Ibid., 96–97.
80. Ibid., 97.
81. Ibid., 96.
82. Ibid., 100–101.
83. Ibid.
84. Ibid.
85. Ibid., 101–102.
86. Frantz Fanon, *Black Skin, White Masks* (London: Pluto Press, 1986), 111.
87. Ibid., 111. Fanon is quoting here from Jean Lhermitte, *L'Image de Notre Corps* (Paris: Nouvelle Bevue critique, 1939), 17.
88. Ibid., 112.
89. Ibid., 109.
90. Ibid., 116.
91. Ibid., 111.
92. Catherine Malabou, *What Should We Do with Our Brain?,* trans. Sebastian Rand (New York: Fordham University Press, 2008), 5.
93. Ibid., 5. Emphasis added.
94. Ibid., 5.
95. Ibid., 6.
96. Catherine Malabou, *Plasticity at the Dusk of Writing*, trans. Carolyn Shread (New York: Columbia University Press, 2010), 37.
97. See Giacomo Rizzolatti and Corrado Sinigaglia, *Mirrors in the Brain: How Our Minds Share Actions and Emotions* (Oxford: Oxford University Press, 2008).
98. See Ruth Leys, "Both of Us Disgusted in My Insula: Mirror Neuron Theory and Emotional Empathy," *Nonsite.org* 5 (March 18, 2012); Ruth Leys, "The Turn to Affect: A Critique," *Critical Inquiry* 37, no. 3 (2011): 434–72; Jaak Panksepp and Lucy Biven, *The Archaeology of Mind: Neuroevolutionary Origins of Human Emotions* (New York: W. W. Norton & Company, 2012). Joseph LeDoux, *The Emotional Brain* (New York: Simon & Schuster, 1996).

BIBLIOGRAPHY

Derrida, Jacques. *On Touching: Jean-Luc Nancy*. Translated by Christine Izarry. Palo Alto, CA: Stanford University Press, 2005.

Fanon, Frantz. *Black Skin, White Masks*. London: Pluto Press, 1986.

Herder, Johann Gottfried. *Herder: Philosophical Writings*. Cambridge: Cambridge University Press, 2002.

Husserl, Edmund. *Cartesian Meditations*. Translated by Dorion Cairns. The Hague: Martinus Nijhoff, 1960.

LeDoux, Joseph. *The Emotional Brain*. New York: Simon & Schuster, 1996.

Leys, Ruth. " 'Both of Us Disgusted in My Insula': Mirror Neuron Theory and Emotional Empathy." *Nonsite.org* 5 (March 18, 2012).

———. "The Turn to Affect: A Critique." *Critical Inquiry* 37, no. 3 (2011): 434–72.

Lhermitte, Jean. *L'Image de Notre Corps*. Paris: Nouvelle Bevue Critique, 1939.

Lipps, Theodor. "Das Wissen Von Fremden Ichen." In *Leitfaden der Psychologie*. Leipzig: W. Engelmann, 1906.

Malabou, Catherine. *What Should We Do with Our Brain?* Translated by Sebastian Rand. New York: Fordham University Press, 2008.

———. *Plasticity at the Dusk of Writing*. Translated by Carolyn Shread. New York: Columbia University Press, 2010.

Panksepp, Jaak, and Lucy Biven. *The Archaeology of Mind: Neuroevolutionary Origins of Human Emotions*. New York: W. W. Norton & Company, 2012.

Péraldi, François. "The Crane-Child." In *Psychoanalysis, Creativity, and Literature: A French-American Inquiry*, edited by Alan Roland, 96–102. New York: Columbia University, 1978.

Rizzolatti, Giacomo, and Corrado Sinigaglia. *Mirrors in the Brain: How Our Minds Share Actions and Emotions*. Oxford: Oxford University Press, 2008.

Scheler, Max Ferdinand. *The Nature of Sympathy*. Translated by Peter Heath. Piscataway, NJ: Transaction Publishers, 1970.

Scherner, Karl Albert. *Das Leben des Traums*. Berlin: Schindler (1861).

Stein, Saint Edith. *On the Problem of Empathy*. Translated by Waltraut Stein. Washington, DC: ICS Publications, 1989.

Vischer, Robert. 1873. "On the Optical Sense of Form: A Contribution to Aesthetics." In *Empathy, Form, and Space*, edited by H. F. Mallgrave and Eleftherios Ikonomou, 89–123. Santa Monica, CA: The Getty Center for the History of Art and the Humanities, 1994.

Chapter 10

Event, Plasticity, and Mutation

Harnessing the Work of Malabou and Badiou in Support of a Molecular Event

Nancy D. Nisbet

Plasticity is "the process by which a contingent event, or accident, touches at the heart of the system." —Catherine Malabou[1]

Of growing interest in continental philosophy is the development of a materialist ontology that embraces meaningful linkages to the natural sciences all the while maintaining commitments to the autonomous subject. The philosophical development of such an ontology faces two difficulties: avoiding the charge of ungrounded metaphysics and resisting the risk of the reduction of the subject to nothing more than scientific laws. In close engagement with epigenetics and neuroscience, Catherine Malabou establishes an ontological concept of plasticity and epigenesis. Remaining at issue in her compelling philosophical account for the immanence of subjectivity within neural matter is the problem of a formal grounding that resists reduction. Alain Badiou's axiomatic ontology, on the other hand, provides a formal mathematical foundation and offers resistance to the risk of reduction. The difficulty here appears to be its application in the realm of biology. In this chapter, I bring the works of Malabou and Badiou together to demonstrate how their work may be productively harnessed to overcome these challenges. Further, I mobilize this philosophical foothold to develop the possibility for considering biological events at the level of the molecule. Through this investigation I develop the possibility for thinking genetic mutation as a molecular event and propose the possibility for a new molecular subjectivity that has the potential to dissolve the distinction between the living and the nonliving.

In order to pursue this inquiry, a philosophical scaffolding must first be constructed. In part I of this chapter, I make six moves as I tack between the distinctive works of Malabou and Badiou. The first move identifies the starting point: genetics, epigenetics, and mutation. Malabou's plasticity, particularly her important consideration of its destructive aspect, marks the direction of the second move. With this introduction of absence as a material form, the third turn is toward Badiou's event and the unlocalizable and unpresentable point of possibility that is the figure of the void. The fourth tack returns to Malabou's evental plasticity; particularly its disloyalty and reprogramming of the biological program itself. The fifth turn introduces Badiou's distinction between being and appearing and the development of the concept of inexistence: the possibility of something to have being without existing. The sixth and final move addresses the problem of contingency, immanence, and indeterminability: can something be included and simultaneously contingent?

With the philosophical rigging in place, in the second half of this chapter I turn to the question of the eventality of genetic mutation. The first consideration here is the high fidelity of genetic material and the contingent occurrence of mutation. The second is to examine the appearing of mutation: how can something new come to appear in this molecular situation? Third, exploration of the range possible consequences of mutation—from the imperceptible to the catastrophic—leads to the question of whether any mutational result be considered evental? It is here that I introduce molecular subjectivity as the material assembly that is instrumental in maintaining the result of the molecular event. The final step in this inquiry considers the range of levels of relation on which mutation operates and emphasizes the difference in the consequence of events at the level of the molecule.

PART I: BUILDING THE SCAFFOLD

Genetics, Epigenetics, Plasticity, and Mutation

With its high degree of fidelity and consistency, the genetic system provides fundamental biological information that underpins the development of each living being. However, in the past thirty years or so it has become evident that the genetic sequence is neither the only nor the principal cause of the vast range of variation observed between species or between individuals of a particular species. The historically dominant belief that the linear sequence of genetic material (i.e., DNA or RNA) determines (like a blueprint or code) all traits and variation has been shown to be false. Meaning upon, over, or above genetics, *epigenetics* is the study of the variable nature of environmental and temporal regulation of expression that goes beyond the linear genetic sequence. Today, it is accepted that most variation is a result of complex and far-reaching epigenetic processes.

In her deep interrogation and development of the concept of plasticity, Malabou has created a remarkable philosophical apparatus that offers support for the emergence of immanent change that is secured by contemporary epigenetics and neuroscience. As a giver, receiver, and destroyer of form, Malabou's plasticity is a process whose results are unpredictable—a process of unforeseen and emergent determination. It is a capacity for excess that is accommodated within the program itself. In the context of neural epigenetics Malabou writes that "plasticity forms where DNA no longer writes."[2] Epigenetic activity, as evidenced in neural plasticity, exposes the discontinuity between the genetic code and the myriad variation that appears. It is this plasticity, above and beyond genetics, that has become the focus of a great deal of ongoing experimental and theoretical research. So why do I reintroduce the question of genetics and its relationship to change? It is because of the undeniable fact that despite its impressive consistency, the genetic sequence can and does undergo unpredictable change—it too is plastic. In each and every living being mutations occur intermittently throughout the course of a life.

The word "mutation" comes from the Latin verb *mutare* meaning *to change*. In his 1859 *On the Origin of Species*, Charles Darwin uses "mutation" to indicate commonplace changes in form or quality,[3] whereas unexpected changes in the appearance of an individual were referred to as "variations." It was not until 1901 in Hugo de Vries *The Mutation Theory* that the word "mutation" took on a biological signification. Today "mutation" is largely a technical term and is defined as "any change in the sequence of the DNA in a chromosome."[4] It is important to emphasize that unrestricted mutation would be catastrophic to the maintenance of life processes. However, despite the abundant and efficient repair mechanisms that resist their appearance, some mutations escape repair and ultimately become indistinguishable as errors—they are incorporated and serve as the template during the next round of replication. With numerous modes of presentation and a wide range of effects, it is clear that mutation is a complex and problematic phenomenon. While it may be possible to reduce the frequency of mutations (by altering our behaviors, for example), there is no way to entirely prevent the occurrence of mutation—sooner or later mutation is certain to arrive.

We might ask why there are mutations at all; isn't the absence of mutation simpler? It is certainly conceivable that replication of genetic material could be error free. The question of mutation is a variant of the ancient philosophical question: *why is there change rather than no change*? Mutation variously describes a result, a process, or a capacity for change. As there is between being and beings, there remains an ontological difference between mutation and mutations. Mutation *qua* mutation is not reducible to its result: neither is it reducible to its cause or process nor to the capacity of genetic material to be changed. It is for this reason that the question of ontological possibilities of thinking mutation as an event becomes relevant. To begin to answer this question I turn to Catherine Malabou's development of the concept of plasticity.

Plasticity and a Form of Absence

In materials science plasticity is notably different from flexibility and elasticity. While "flexibility" is a term that simply refers to matter's capacity to bend,[5] elasticity designates the capacity for matter to return to its initial form after deformation[6]. Broadly, these terms describe a material's ability to absorb change and maintain stability. "Plasticity," on the other hand, is a term used to describe the persistent and irreversible physical deformation that results from excess force. In addition to its signification in giving and receiving of form,[7] Malabou further insists on the importance of acknowledging plasticity's destructive potential. In both *The New Wounded* and *Ontology of the Accident*, she develops the concept of destructive plasticity and discerns the appearance of a new and discontinuous subjectivity that is not an absence of form, but rather a *material form of absence*.[8]

This being of absence (an absence that *is*) recalls to mind the dialectic of being and nothing. In the *Science of Logic* Hegel states that pure being and pure nothing are the same. He goes on to clarify that they are not the same—they are distinct but nevertheless "equally *unseparated and inseparable*."[9] Hegel continues that "nowhere . . . is there anything which does not contain both being and nothing in itself."[10] Malabou echoes this and writes, "Every form carries within itself its own contradiction."[11] It is not difficult to link the creation of this absence through destructive plasticity to Hegel's assertion that being includes the *determination of the negative* such that "activity, creation, power, and so forth, are the bringing forth of an other."[12] Said another way, there is always something included, something at once unseparated and inseparable, that makes possible something radically different. This emergence of discontinuous change brings us to the nature of the event.

The Event and the Void

For Alain Badiou the event is a contingent discontinuity. It is a singular change that appears as an excess that ruptures what came before: it cleaves the laws of appearing in a world and makes possible the appearing of something radically new. He emphasizes that an event is "the creation of a new possibility": in addition to changing the real, it changes what is possible—the very "possibility of possibility."[13] Again, this calls to mind the problematic gap between being and beings. Is there a way to think across this divide? In *Being and Event* Badiou insists on some initial presentation of being that is already-there: "where is the absolutely initial point of being? Which initial multiple has its existence ensured such that the separating function of language can operate therein?"[14] Badiou's approach to circumventing this ontological impasse takes advantage of the generic relations of belonging that characterize axiomatic set theory and its capacity to speak and think about *that which is without a determinate concept.*

Zermelo's axiom[15] is a mathematical demonstration that there always exists some Other, an indistinguishable excess, that is contained within every existing multiple. The empty set is this excess multiple that is included in all multiples. It is in this way that Badiou identifies "the necessary figure that designates the gap"[16]—*nothing*—it is the name of the void. The empty set {ø} is that indistinguishable multiple that contains no elements and is the unique unpresented form of presentation itself. In having no elements that belong to it, the empty set is unable to "present itself in its difference" and is, therefore, subtractive—it negates presentation.[17] Badiou writes that the void is "the first multiple, the very beginning from which any multiple presentation, when presented, is woven and numbered."[18] In keeping with Aristotle's demonstrations concerning the void, Badiou observes that it is indifferent, infinite, and unmeasured.[19] In contrast however, Badiou contends that *the void is*. It is an unrepresentable and unlocalizable "punctuality"—the void is "the unpresentable *point of being*."[20]

Echoing Hegel on this point, Badiou writes, "The void is indeed included in everything."[21] Ontologically, the void is that unique existent multiple—we might even say that *form of absence*—that is a finite originary limit of every multiple.[22] Badiou refers to this lower limit as an *evental site* which "can be said to be 'on the edge of the void.' "[23] It is precisely because there is *nothing* beneath the empty set that the evental site is foundational. In this way, Badiou's axiomatic ontology provides a means to avoid the charge of ungrounded metaphysics.

Disloyalty and the Eventality of Plasticity

Evental plasticity is something like a contingent possibility of disloyalty that haunts the program from within. As Malabou describes it, destructive plasticity is "an agency of disobedience to every constituted form, a refusal to submit to a model."[24] It is not only an infidelity to the usual functioning of the system, but moreover, a loyalty to this disloyalty that is inescapable. In his thinking of the event, Badiou also employs the trope of fidelity. He argues that the effects of an event may only be sutured to appearing in the world through the active fidelity of a subject.[25] The essential requirement of the subject is fidelity toward the *consequences of the event* rather than to *what there is* in a world. For Badiou, this *what there is* is nothing other than what belongs (i.e., what is familiar and known): it is the usual functioning of the system. We could say that the rupturing excess of evental change only appears in a world through faithful disobedience: a loyalty to the rupturing possibility of radical change and new possibility, and a local disloyalty to established structure.

Malabou explains that plasticity can be evental precisely because it is able to "eventalize the program" by deprogramming it.[26] An event's occurring is always contingent, but the possibility that it can occur is inscribed in its own law—a law that allows for the excess of its own infringement—and yet "does not allow us to anticipate its instances."[27] Similar to the figure of the void that marks the unpresentable point of being discussed in the previous section, plasticity reveals an internal excess; an indiscernible inclusion of something supplemental that is both a contingent possibility and an unanticipated occurrence. As she writes in the *Ontology of the Accident*, "Destructive plasticity enables the appearance or formation of alterity where the other is absolutely lacking."[28]

In parallel with Malabou, Badiou confirms that what we are looking for is "an indiscernible internal to a situation. . . . We want the set to exist in the very place in which it is indiscernible."[29] The difficulty that we seem to encounter here is that of the simultaneity of inclusion, indiscernibility and contingency. How can something be internal to the situation and, at the same time, indistinguishable? Further, how can it be, at once, indistinguishably at the heart of the system and yet contingent? Badiou's theory of inexistence opens the door to a possible solution.

Inexistence and Indiscernibility

Rather than the traditional philosophical distinction between being and existing, Badiou crucially distinguishes between being and appearing. Similar to Heidegger's concept of *dasein* (but without its restriction to human being), existence is a localization of a being-there. It is a category of appearing in a world, rather than a category of being.[30] According to Badiou, appearing is a being's logical relation of identity to itself and to other beings. So, unlike the ontological situation where a multiple either is, or is not, identical with itself; a being's self-identity is variable "to the extent that it appears in the world."[31] This is to say that something in the world can exist more or less—it can even inexist.[32] In this way, something can be indiscernible: it can have *being without existing*: its existence is inexistence.

To be clear, inexistence is not the same thing as nonbeing: *inexistence is*. The ontological argument for the being of inexistence begins, as earlier discussed, with the mathematical proof of the inclusion of the empty set in all multiples. The question now becomes, how are we to account for this empty set at the level of appearing? In order to think inexistence in a world, Badiou specifies that there must first be a reference multiple that can be affirmed not only as being, but also as appearing in a world. He explains, "the formal precaution always tells us to inscribe appearance in a world under the sign of a referential multiple, guaranteed to be an element (in the ontological sense) of the world in question" this referent "is at bottom nothing but a guarantee of being-in-the-world for the appearing of its elements."[33] With the assurance that a given multiple appears in a world, it is therefore possible to think the empty set therein. So, with the inexistence of an evental site within anything that appears in a world, it is possible to account for something indiscernibly included in a situation. But what of the problem of contingency and immanence? Is it possible for something to be in some way included and simultaneously contingent?

Contingency and Immanent Possibility

The term "contingency" comes from the Latin verb *contingere* meaning to happen and it specifies that something could be otherwise. Contingency further resolves into two significations: one refers to something that has *the potential to appear*; the other refers to *the being of the potential* itself. In this sense, contingency can be said to be subtracted from the being (or nonbeing) of any potential thing. The pertinent question to ask is whether there can there be something like an endogenous event.

In Leibniz's idea for a *characteristica universalis*, or universal language of thought, Badiou notes a parallel problem for contingency. This "complete language"[34] implies intrinsic nomination but, for Badiou, the world is not a complete language into which "being is folded."[35] He argues that the contingent event is rejected by the logic of Leibniz since "if being is coextensive with a complete language [for Leibniz], it is because it is submitted to *intrinsic denominations*, and not to an errancy in which it would be tied to a name by the effect of a wager."[36] This denomination is the setting apart of a thing through language, specifically, by a name. The problem is that if being is entirely submitted to this differentiating power of language, language would have already internally set apart all possible presentations and, by definition, being would be closed to the event. We are left with the dilemma that events do seem to happen.

It is for this reason that Badiou makes this notable statement of exception: "There are only bodies and languages, except that there are truths."[37] While there is not space in this chapter to delve into his unique concept of truth, it serves to point to the contingent disruption of a system and the emergence of something new. A truth is created within a world, but it is an excess to the system of that world. For Badiou, truths are not "what there is" in a world, but rather, truths are discontinuous ruptures of what there is—they are *what comes to be*.[38] These truths concern neither knowledge nor meaning: a truth is no more than the appearing of a new creation that does not make sense from the perspective of what there is in a situation. For Badiou, "senselessness [l'insensé] is the primordial attribute of the True."[39]

The apparent difficulty in accounting for contingency arises when we think the problem beginning with the concept of a complete program. Consider the reverse: that the differentiating process of intrinsic denomination is not the same as—and does not imply—a completely determined system. As explained earlier, there exists a possibility to axiomatically set something apart without thereby giving it a proper name. In its indication of a scheme of relationships, an axiom is a form of generic instruction—a program of sorts—that retains an element of indetermination such that it does not specify any particular possibility or instance: axiomatization overcomes the impasse of the name.

Starting with the figure of the void as the foundational gap of inexistence that is universally included in all beings, and continuing with the possibility of axiomatically avoiding the impasse of internally discerning a determinate name, the problem of inclusion and indiscernibility is averted. The evental site can be both unrepresentable and unlocalizable. The resolution of contingency into the potential of something appearing and the potential for this possibility itself allows the event to remain endogenous and unnecessary. An event's possibility is both internally unpresented as the inclusion of the figure of inexistence, and entirely contingent as the evental site does not guarantee the occurrence of an event, nor that an inexistent will come to fully exist—the site marks the endogenous gap of contingent possibility. What is accommodated in the system is only *the possibility of the errant*, not any particular errant result. It is thus possible to think a program that is open to (and includes) the contingency of the event and thereby find resistance to scientific reduction.

PART II: GENETIC MUTATION AND THE POSSIBILITY OF A MOLECULAR EVENT

Biological Systems and the Situation of Mutation

Biological systems are porous and responsive to the environment; however, in the sense of rejecting transcendental causation, they are also closed. This closure is not to say that such a system is fully determined. As Badiou explains, it is necessary that "every world be ontologically infinite": no world is reducible to its component parts.[40] He clarifies that the infinity of a world is of the "inaccessible" kind—it is "nontotalizable."[41] There is, as Badiou notes, a paradoxical "operational closure" of a world that is simultaneous with its "immanent opening."[42] Malabou recognizes a similar situation in the world of contemporary neurobiology where plasticity describes an indeterminate and modifiable network of neural connections. Rather than some sort of predetermined computing biomachinery, the brain functions as a highly malleable and openly responsive biological system that is, in some manners, pervious to the environment. But this plasticity is more than a homeostatic mechanism that fluctuates within prescribed limits—it is also an energetic process of creating entirely new structural forms. As she writes, it is "a mutation 'right at the level of form.'"[43] Plasticity is precisely the infinite "possibility of a closed system to welcome new phenomena."[44]

Let us return to genetic mutation. Recall that mutation *qua* mutation is not reducible to its process, its result, or to a capacity. There exists the familiar ontological gap between its being and its presentation. Badiou's *evental site*—the figure of the void as the unpresentation of presentation itself—provides access to thinking across this gap. As is well known, the accurate repetition of the genetic sequence is the standard situation: it is a high fidelity program of sorts. What results is to be an exact copy of what came before. It is, however, also the case that mutations occur; they sometimes escape repair and become integrated in the sequence. Similar to Malabou's description of plasticity above, with mutation the genetic program also reveals the constitutive possibility of disobedience. That a system can be functionally closed and yet contingently open to infinite possibility is also at the center of the question of evental mutation.

Mutation is contingent and something that is both indistinguishable and internal to the genetic situation. Badiou's concept of inexistence becomes helpful in understanding how this is possible. Recall that, as inexistent, something can have being without existing: inexistence *is*. In *Logics of Worlds* he demonstrates that "a real point of inexistence is traced in existence."[45] In every object that appears in a world, this inexistent point marks the contingency of being-there. In the context of mutation, we have two broad regimes of appearance: global (e.g., the cell or organism) and local (e.g., a particular molecule of DNA in one cell). The appearing of this local sequence of DNA therefore guarantees the inclusion of an evental site and the inexistence of mutation.

The Appearing of Mutation

Developed in his *Logics of Worlds*, Badiou's theory of appearing is based on his concept of transcendental indexing which signifies an order structure of a world. It describes the "network of identities and differences that concern the elements of what appears."[46] To be clear, this transcendental is not the

Kantian transcendental: it does not make anything appear or otherwise act in distinction to a world. For Badiou, the transcendental of a world is a mathematically based vectorial method by which one can guarantee the being-there of an object in the world. Based on the internal organization of elements of a multiple, and the relational indexing of this multiple to another multiple in a world, something can come to appear as an object in a world. A transcendental index exists between any (and every) pair of beings in a world. It is this multiplicity of relations that accounts for the possibility of something's appearance as a new unity; these possible alternate unities transcend—this is to say, exceed—the beings that are there.

While necessarily the main topic of a future essay, it is worth noting that Malabou also rethinks the Kantian transcendental and thereby accommodate the appearance of discontinuous change from within. In her recent book, *Before Tomorrow*, she develops an epigenetic paradigm in which the transcendental can be accounted for from within the world as a surface contact point (like a seismic epicenter) of effect. She argues that the value of the transcendental is vectorial: it is "a passage and conductor between invariance and modification."[47] Under this new paradigm, the transcendental is lawfully unstable and contingent. As she concludes in an earlier essay, life "modifies the transcendental": the transcendental is plastic.[48]

What can we say about the appearing of mutation? By way of Badiou's argument for the being of inexistence and the universal inclusion of an evental site, we can say that there is a founding other that has a minimal intensity of appearing within the genetic material of a cell. I call this the *mutational site*. Prior to its appearing mutation is said to inexist. On condition of an event, the identity of this element with itself becomes maximal and a mutation comes to fully exist as a new point in a world—a specific mutation has been "worlded" in the cell.[49]

How shall we evaluate its degree of relation to the DNA in another cell of the same organism? Remember, with mutation the normally accurate process of replication is interrupted: something new comes to appear. The former repetition is cut and an entirely new sequence—discontinuous with the first—begins. This is not to say that there are no similarities between the former and the latter identity: they are identical—except for the mutation itself. However, at the precise location of the mutation, the two identities have nothing at all in common: they are completely different. Thus, according to some transcendental lines, there may be a high degree of identity (but not maximal) between them—most of the sequence remains identical. Along a different transcendental, say the transcendental that specifically locates the mutation, there will be minimal identity between them. Badiou clarifies "that what governs appearing is not the ontological composition of a particular being (a multiple) but rather the relational evaluations which are determined by the situation and which localize that being within it."[50] In this way, we can account for the localization and appearance of a mutation in this molecular world. The question remains whether such a mutation can be considered an event. Would it always be evental?

Evental Mutation

With respect to the understanding of event put forward by Badiou we can now consider the soundness of the concept of evental mutation. Of mutation we can confirm that it is a discontinuity that—in its explosive appearing and disappearing—creates the possibility for something inexistent to come to full existence in a world. First there was no mutation—mutation is nothing—then, suddenly, it appears. This discontinuity is a real cut with the former sequence of the repeating count. The relation across the gap that forever separates them is one of nonrelation. This cell, with the irruption of its new genetic sequence, marks the commencement of a new lineage. It is precisely because of these similarities that we can acknowledge the possibility that mutation is a molecular event.

We must now consider that mutations almost always disappear through the efficient actions of the aforementioned molecular repair mechanisms. It is difficult to say whether a transient situation of mutation can be considered evental. With the disappearance of a mutation through repair, it seems possible to say that there was no event. Nothing happened—except that something did happen—if only for a moment. Recall, in Badiou's account there is no guarantee that the occurring of an event will result in a real change. The event is only the initial possibility for something inexistent to come to fully exist—what is also necessary is the loyalty of a subject. In the case of a repaired mutation it is, perhaps, more accurate to say that there was nothing that held to the consequences of this event; there was no loyal subject. The evental consequences evaporate with the initiation of molecular repair. Even though this repaired mutation ultimately retains its status as inexistent, the possibility for it to have existed really did appear.

Another case to consider is one in which a mutation escapes the vigilant molecular guardians of repair: its status as an event is more convincing. Being unrepaired, the mutation remains and appears as an element in a particular cellular multiple. I propose that a *molecular subject* retains the consequences of the mutational event and the mutation persists. Consider, when repair is evaded, the usual functioning of the molecular repair apparatus (an assembly of molecules) is somewhat different. Mutations are not an easy fit. Something like an intervention—a subjective molecular commitment that is disloyal to the program—is needed to transform the trace of the mutation into its possible inscription. It is in this altered functioning that we can say this transient molecular subject holds to the trace of the event long enough for the mutation to become inscribed. On condition of an evental mutation, the genetic sequence—the ordered elements that make up this multiple—has forever changed. Remaining in place by evading repair, the mutation has, in one sense, come to belong in the world. From the perspective of the cellular lineage, the errancy of this incorporated mutation disappears.

It must be asked whether it is legitimate to think the subject in the molecular world of mutation? In the unavoidably brief sketch that follows, I argue that this speculative idea is consistent. Both Badiou and Malabou, in different ways, importantly unhinge the subject from human consciousness and radically open the range of possibility of thinking change in a world. For Badiou, the subject "stands at the crossing between a lack of being and a destruction, a repetition and an interruption, a placement and an excess."[51] For Badiou, there is no requirement for human consciousness or experience, there is not even a requirement for a subject to be human. His subject is fundamentally an agency of change that participates in the creation of something new: a "relation between an event and a world."[52] For Malabou, the subject is neither an origin nor a result, but a plastic process which is open to indeterminate and indifferent change. Her development of the neuronal connection to subjectivity[53] presents a very interesting possibility for a molecular conception of a subject. If the subject can, in fact, be said to be continuous with the material and plastically creative brain, it is not so difficult to argue for molecular subjectivity in the world of living cells.

Let us return to the case of a mutation that has escaped repair and is genetically inscribed. It is important to note that all persisting mutations are not equal. The consequences of a mutation range from minimal to catastrophic. On the one hand, a mutation may have no effect whatsoever. For a variety of reasons, a mutation may also be inconsequential: the mutation is present but the cell is able to continue to function exactly as required. The mutation may also be latent: it may have no immediate effect, but over time, or with the possibility of the appearing of other mutations, there may be functional consequences. On the other hand, the mutation may have immediate consequences to the cell's functioning, even to the point of being catastrophic. Similar to the case of immediate repair, in this catastrophic appearance of mutation, it is once again difficult to decide whether the mutation was evental. There was one consequence—cell death. Repetition is cut, but it is more than a cut, it is extinguished.

It is important to remember that mutation occurs under several levels of relation within an organism. Evental mutation happens in molecular genetic material of living beings. Its immediate result is a molecular change in the genetic sequence. At the molecular level, the cell is indifferent to the event. This is not to say that the cell cannot perceive the change, or that the change does not produce any consequences; rather, the cell continues to function according to its capacity under the mutational effects—whatever they may be. At the level of the organism, however, things may be quite different. There can be a significant delay in the perception of any noticeable effects of an evental mutation. The real-time occurrence of a mutation is imperceptible to complex multicellular organisms. The mutation that has occurred and contingently becomes existent for the cell, remains inexistent for some time (perhaps indefinitely) from the perspective of organism. It is only through various processes (cellular proliferation, additional mutations, depletion or excess of biochemical products, etc.) which lead to a dramatic magnification or intensification of effects that an evental mutation can be perceived at the level of the organism. The situation of evental mutation is different for the multiple that is the organism than for the multiple that is the cell: in agreement with Badiou, it can be said that "the only truth is a local one."[54] What is local to a cell is not necessarily local to the organism.

CONCLUSION

These molecular events and molecular subjects are of particular interest going forward. While they are classified as organic or inorganic, molecules are not the kinds of things that can be classified as living or nonliving. Above the level of the molecule itself, all matter—whether a grain of sand, a drop of water, a bacterium, or a human being—is composed of molecules. In this way, molecules already dissolve the living/nonliving distinction. It is for this reason that molecular events have the potential to serve in a new materialist ontology. Further, as aconscious entities that are instrumental in effecting unprogrammed change, further development of the concept of molecular subjectivity may offer insight into questions of consciousness and the autonomous material subject.

In this short chapter, I used the work of Malabou and Badiou to construct a philosophical scaffolding that resists the charge of ungrounded metaphysics and avoids reduction to scientific law. This architecture supports the development and affirmation of the possibility of evental mutation and a new molecular subjectivity. I have shown that genetic mechanisms are both more, and less, than programs determining the repetition of life. On one hand, there is a high degree of fidelity to the replication of genetic material in living beings, both within the span of an individual life and across generations. On the other hand, there is an internal errancy to the functioning of these mechanisms. Paradoxically, mutation is—and is not—subject to chance. Its specific location, type, and timing are all contingent, but the possibility of mutation is not: mutation is certain to happen sooner or later. Contingent molecular events are not just tolerated, but are inextricably included within the program.

In addition to the molecular event, I also demonstrated the possibility for a new subjective form: the transient molecular subject is that material assembly which maintains the effects of the mutational event against the rule of repair. In the profoundly stable world of molecular genetics, evental mutation and the molecular subject create the fundamental possibility for immanent and unscripted change. With the appearance of mutation genetics demonstrates its own plasticity: genetic material includes a space of indeterminacy that conditions the possibility of possibility.

Living organisms are not unified wholes but, rather, multiplicities of systems and frontiers in dynamic relation. Neither unresponsive nor isolated, these homeostatic biosystems are plastic and mutable. From the level of the molecule to that of the entire organism, living systems exhibit remarkable balance, and even while they maintain this precarious tension, they are susceptible to rupture. From the larger complex and sophisticated neural program to the fundamentally ubiquitous and largely conserved basic programming of the genetic code, the contingent, indifferent, and interrupting effects of the event is evident in living beings. In place of reduction, the complexity and unlawfulness of molecular plasticity is confirmed.

> "Only in making explosives does life give shape to its own freedom."[55]
>
> —Catherine Malabou

NOTES

1. Catherine Malabou, *The Future of Hegel: Plasticity, Temporality and Dialectic*, trans. Lisbeth During (London: Routledge, 2004), 192–93.
2. Catherine Malabou, *Plasticity at the Dusk of Writing: Dialectic, Destruction, Deconstruction*, trans. Carolyn Shread (New York: Columbia University Press, 2009), 60.
3. Charles Darwin, *On the Origin of Species by Means of Natural Selection*, 1998. Project Gutenberg, 219, 404, 503, 520.
4. Paul Griffiths and Karola Stotz, *Genetics and Philosophy: An Introduction* (Cambridge: Cambridge University Press, 2013), 21.
5. "flexibility, n." OED Online. September 2016. Oxford University Press. http://oed.com/view/Entry/71523?redirectedFrom=flexibility. (Accessed October 18, 2016).
6. W. D. Callister and D. F. Rethwisch, *Materials Science and Engineering: An Introduction* (Hoboken, NJ: Wiley, 2007), 138.
7. Catherine Malabou, *What Should We Do with Our Brain?*, trans. Sebastian Rand (New York: Fordham University Press, 2008), 5.
8. Catherine Malabou, *The New Wounded: From Neurosis to Brain Damage*, trans. Steven Miller (New York: Fordham University Press, 2012), 18.
9. Georg Wilhelm Friedrich Hegel and George Di Giovanni, *The Science of Logic* (Cambridge: Cambridge University Press, 2010), 60.

10. Ibid., 61.
11. Malabou, *What Should We Do?*, 71.
12. Hegel, *The Science of Logic*, 62.
13. Badiou, “Is the Word ‘Communism’ Forever Doomed?” Lecture Transcript. *Lacanian Ink 32*, 2014. np.
14. Badiou, *Being and Event* (London: Bloomsbury Academic, 2007), 48.
15. Ibid., 186. Zermelo’s Axiom as written in is as follows: $(\forall\alpha)\ [(\alpha \neq \emptyset) \rightarrow (\exists\beta)\ [(\beta \in \alpha)\ \&\ (\beta \cap \alpha = \emptyset)]]$. This can be read: for every nonvoid set α there exists β such that β is an element of α, and such that the intersection of α and β is void.
16. Ibid., 55.
17. Ibid., 67.
18. Ibid., 59.
19. Ibid., 73.
20. Ibid., 77.
21. Ibid., 87.
22. Ibid., 186.
23. Ibid., 175.
24. Malabou, *What Should We Do?*, 6.
25. It is important to note that the subject is of paramount importance to Badiou’s philosophy and the subject of his book *Theory of the Subject*. Detailed examination of his concept of the subject exceeds the scope of this chapter.
26. Malabou, *What Should We Do?*, 8.
27. Catherine Malabou, *The Ontology of the Accident: An Essay on Destructive Plasticity* (New York: Polity, 2012), 30.
28. Ibid., 11.
29. Badiou, *Being and Event*, 375.
30. Alain Badiou, *Second Manifesto for Philosophy* (Cambridge: Polity, 2011), 57.
31. Alain Badiou, *Logics of Worlds: Being and Event II* (London: Continuum, 2013), 246.
32. Badiou, *Second Manifesto*, 57.
33. Badiou, *Logics of Worlds*, 245.
34. Badiou, *Being and Event*, 320.
35. Ibid., 317.
36. Ibid., 319. Emphasis added.
37. Badiou, *Logics of Worlds*, 569.
38. Alain Badiou and Peter Hallward, *Ethics: An Essay on the Understanding of Evil* (London: Verso, 2012), 25. Emphasis added.
39. Badiou, *Logics of Worlds*, 522.
40. Ibid., 309.
41. Ibid., 309.
42. Badiou, *Logics of Worlds*, 310.
43. Malabou, *The Future of Hegel*, 192.
44. Ibid., 192–93.
45. Badiou, *Logics of Worlds*, 322.
46. Ibid., 596.

47. Catherine Malabou, *Before Tomorrow: Epigenesis and Rationality*, trans. Carolyn Shread (Cambridge: Polity, 2016), 183.

48. Malabou, "Can We Relinquish the Transcendental." Author's manuscript, 16.

49. Ibid., 114.

50. Ibid., 156.

51. Alain Badiou, *Theory of the Subject* (London: Bloomsbury Academic, 2013), 139.

52. Badiou, *Logics of Worlds*, 79.

53. For example, in *The New Wounded* and Adrian Johnston and Catherine Malabou, *Self and Emotional Life: Philosophy, Psychoanalysis, and Neuroscience* (New York: Columbia University Press, 2015).

54. Badiou, *Logics of Worlds*, 141.

55. Malabou, *What Should We Do?*, 73.

BIBLIOGRAPHY

Badiou, Alain. *Being and Event*. Translated by Oliver Feltham. London: Continuum, 2007.

———. *Second Manifesto for Philosophy*. Translated by Louise Burchill. Cambridge: Polity, 2011.

———. "Is the Word "Communism" Forever Doomed?" Lecture Transcription. *Lacanian Ink* 32, 2014.

———. *Logics of Worlds: Being and Event II*. London: Bloomsbury Academic, 2013.

———. *Theory of the Subject*. Translated by Bruno Bosteels. London: Continuum, 2009.

Badiou, Alain, and Peter Hallward. *Ethics: An Essay on the Understanding of Evil*. London: Verso, 2012.

Callister, W. D. *Materials Science and Engineering: An Introduction*. 7th ed. New York: Wiley, 2007.

Darwin, Charles. *On the Origin of Species*. Project Gutenberg e-book, 1998. (Accessed February 19, 2014).

Griffiths, Paul, and Karola Stotz. *Genetics and Philosophy: An Introduction*. Cambridge: Cambridge University Press, 2013.

Hegel, Georg Wilhelm Friedrich, and George Di Giovanni. *The Science of Logic*. Cambridge: Cambridge University Press, 2010.

Johnston, Adrian, and Catherine Malabou. *Self and Emotional Life: Philosophy, Psychoanalysis, and Neuroscience*. New York: Columbia University Press, 2013.

Malabou, Catherine. *Before Tomorrow: Epigenesis and Rationality*. Translated by Carolyn Shread. Cambridge: Polity, 2016.

———. "Can We Relinquish the Transcendental." Author's manuscript, 2014.

———. *The Future of Hegel: Plasticity, Temporality and Dialectic*. Translated by Lisabeth During. London: Routledge, 2004.

———. *The New Wounded: From Neurosis to Brain Damage*. Translated by Steven Miller. New York: Fordham University Press, 2012.

———. *The Ontology of the Accident: An Essay on Destructive Plasticity*. Translated by Carolyn Shread. Cambridge: Polity, 2012.

———. *Plasticity at the Dusk of Writing: Dialectic, Destruction, Deconstruction*. Translated by Carolyn Shread. New York: Columbia University Press, 2009.

———. *What Should We Do with Our Brain?* Translated by Sebastian Rand. New York: Fordham University Press, 2008.

Part II

THE POLITICS OF PHILOSOPHY

Chapter 11

Reading Derrida's *Glas*

A Queer Presence alongside Hegel

Michael Washington

An opening onto the text

This reading of *Glas* is placed right at the edge between the questions of what is real[1] about queer versus what is symbolizable about it, in that what Derrida sees in Genet and how that thing touches the Hegel column opens up a triangulation of questions that implicate Writing (or the logic of supplementarity) in the psychically clandestine and underground work of sexuality. Such a writing of sexuality could then be seen to interlock with and interrogate Absolute Knowledge as understood by Derrida. What *Glas* evokes is the possibility of sexuality's submission to the symbolizing function in ways that push that function to its limit, that provoke its terrifying realization that its horizon, the limit point at which it fails, exists and is carried within it as its essential possibility; and this prevision, this anticipation of a writing of the real, a "distyle" that is, presents itself to Derrida as his only chance of escape from Hegelianism altogether.

What is interesting about *Glas* is out of all the figures that Derrida could have placed alongside next to Hegel, he placed Genet, a queer writer. We know, from the first page of the book (a text written in two phallic columns placed next to each other, giving off the effect of having been written at the same time), that the two columns were meant to be "unequal," and written in a failed writing that serves a very important purpose: such a rhetorical gesture of a failed grammar allows for the philosophical thesis of a critique of logocentrism to be felt that much more viscerally. Derrida is able to expose a prose column written on Hegel—*the* "magisterial" figure of a metaphysics of presence of the idea—he is able to open up the logos of Hegel to a destabilizing entry of its other, the antiprose of Genet that "envelop(e)(s) . . . sheath(Es), incalculably reverses, turns inside out, replaces, remarks, overlaps" the Hegel column.[2] A column that Derrida invites us to read as a spectral figure bereft of a father/origin that does nothing but roam and err away from an original point of intention. This is Genet for Derrida, a bastard son of writing that takes a traditional model of kinship and turns it on its head, only after decapitating the head first, such a head being an ideal for Derrida, or a signifier adequate to its signified.

Given all of this thrill of deconstructive play that is found in both columns, what is more interesting is what Derrida finds in the Genet column but does not bring to full articulation. What he first locates in Genet and privileges as an alternative to the autonomous logos associated with Hegel is Writing, a radical supplementarity within the chain of signification that occurs without ever supporting the emergence of a coherent subject; but what is also discovered but less commented upon is the mode through which this deconstructive potentiality is given: a radical queer sexuality, radical mostly in terms of its detonating negative energy.

Whence enters onto the scene that which has been waiting in the wings of *Glas* to be read: what has come to be known as the "antisocial" turn in queer theory, and all of its thoughts on sexuality as understood psychoanalytically. Lee Edelman writes in *No Future: Queer Theory and the Death Drive*, that queerness is what "exposes the obliquity of our relation to what we experience in and as social reality, alerting us to the fantasies structuring in order to sustain it and engaging those fantasies through the figural logics, the linguistic structures, that shape them."[3] He continues on regarding how the fantasy is politicized: it is through those "linguistic structures" that "social reality" is framed, that is to say, what has come to be our social reality is given to us through the forms of fantasies of normativity: fantasies "of an order, an organization, that assures the stability of our identities as subjects and the coherence of the Imaginary totalizations through which those identities appear to us in recognizable form,"[4] an imaginary symbolic interplay constructed to "screen out" and cover over the fact of the "emptiness" that characterizes the "signifier(s)" that constitute and uphold those fantasies.[5] Edelman's way out of the bind of the regulatory nature of the fantasies that order our lives is through insisting on the importance of queerness being lived out at the level of the sexual within the social. Through tapping into the force of sexuality, queerness can explode the regulatory regimes of organizing symbolic fantasies. What is unthinkable for Edelman is to consider sexuality apart from its inherent agency of negation, the "negativity of sex" that constitutes our real "status as . . . nonsovereign subjects."[6] Sexuality, for Edelman, is what affords us a "momentary access to a sense of [one's] radical unrepresentability,"[7] and it is this position that places Edelman squarely within the antisocial company of Genet.

So it is toward a nonrelational Genet that I would like to bend my reading of Derrida, and to begin to do so through posing a philosophical question to *Glas*: when we consider the nature and character of sexuality's negating force that, according to Genet and Edelman, would leave everything decapitated in its wake, drowned in vomit and the refuse of the bodily functions—a negativity that Derrida specifically places next to Hegel as a nonpresence that constantly threatens to penetrate and remark the Hegelian logos—might it still be possible to think of this negating force as not just a transcendental way out of a Hegelian dialectical logic, not just as an alternative to Hegel, but also, as a generative, productive, determining negation, just not determining of new psychic forms that would emerge in presence, but in the mode of the negative, as negative possibilities?

An opening into the text
"Let us start, as always, by quoting":[8]

> It is very arid on the endless esplanade, but it (ça) does nothing but begin, the labor, here, from now on. As soon as it (ça) begins to write. The thing is oblique. It forms an angle, already, with the ground. Slowly bites again [*Remord*] its shadow, dead sure (death) of (it)self . . . they say distyle [*dissent-ils*], if it falls (*ça*) falls (to the tomb) . . . no more than one piece is missing.[9]

Is this "missing" thing that which remains after the fall of Writing? Certainly it is a question of remains, of refuse left behind after the fact, that opens the entire book of *Glas* (both the Genet column and the Hegel one). What remains after the work of writing begins, after this gesture of "distyle"? "Two unequal columns, they say distyle [*disent-ils*], each of which—envelop(e)(s) or sheath(es), incalculably reverses, turns inside out, replaces, remarks, overlaps [*recoupe*] the other."[10] Hinting at a part-answer, Derrida suggests that "of the remain(s)" of writing, there are two different functions, two different possible outcomes: one that would relieve (relevé) the signifier of its anxiety of (after its "fall" into the endless dissemination within the signifying chain) not being able to faithfully represent its place of origin, and would accomplish this by "[relieving] the fall [*chute*] into a monument";[11] the second possibility, a letting go, a letting "the remain(s) fall," as always in *Glas*, to the "tomb(stone),"[12] the death pit of clear univocal sense. The paraph ("the abbreviation of a paragraph: what is written on the side, in the margin")[13] that is stitched next to the description of this second option gives a definition of *catachresis*, Derrida here taking our hand and instructing that this catachresis is the nature of this second kind of response to the consequences of the fall of writing: a "trope wherein a word is diverted from its proper sense," creating a "harsh and unfamiliar dissonance."[14] It is this second option, this antistyle of diverting words from their "proper" sense par-excellence, that Derrida sees in the writing of Genet, and it is for this reason that he (Genet) has been propped up as the alter-column to Hegel in the text. "The remain(s)" of Genet's writing for Derrida is "indescribable . . . rigorously undecidable."[15]

Everything in Genet, according to Derrida—every motif, every trope, conceit, all of the objects that are important to him—he throws away, bends them from their proper sense, tosses them into the pit. We see this in the way Derrida reads these strange objects—flowers, a Vaseline tube, spit—that appear in Genet's texts. To begin with flowers, Derrida comments on their presence in Genet's texts:

> If then, there is no language of flowers, if the flower is in (the) place of zero signification, how can this symbolic zero *take hold* in a jungle of signs and figures belonging to the natural tongue, to nature . . . If flowers figure 'infernal props?' It is because, signifying nothing, they are nonetheless the support of . . . the whole text.[16]

He then goes on to cite from Genet's texts themselves: "You see, but you cannot see, you are necessarily blind to the fact that flowers, not themselves shown, hardly even promised, are constantly being stolen, filched, swiped from you."[17] Flowers here are bound up with signs, that is to say, with "castration" and the "phallus";[18] these "such apparently conventional flowers" are signifiers that take on phallic resonances, are "already worth their weight in sperm and phallus."[19] Whence the description of a sign that stands erect like a flower stem, and if cut off or cut down, opens onto all kinds of problems like anxiety, terror, and the need to invent, to weave together something (anything) that can stand in its place.

The flower/phallic sign that can also be used as a sword opens up a central theme in *Glas*: "antherection," the antierection, the ana-phallus, what contradicts, pushes against the sign that would "band" itself together "erect."[20] The antherection for Derrida is "the time of erection countered, overlapped [*recoupée*] by its contrary—in the place of the flower . . . the overlap goes over itself again indefinitely. Whence this effect of capitalization, but also of unlimited outpouring."[21] Like all of *Glas*, Derrida's invented neologisms come directly out of the speech, the text, of Genet himself. So Derrida locates the operation of antherection, a cutting away from the site that has already been cut away, in the penal colony that the convicts inhabit in *The Thief's Journal*, more precisely in the fact that the place is going to be closed down: "if the colony itself expressly defines a castration, the closing of the penal colony is felt as castration of castration."[22] He goes on to cite Genet: "our most dramatic movement has been cut . . . our exodus, the embarkation, the procession on the sea, which was performed with bowed head. The return, this same progression in reverse, is without meaning. Within me, the destruction of the colony corresponds to a kind of punishment of punishment: I am castrated, I am shorn of my infamy."[23]

To castrate castration. What is an "original" castration? Is it the subsequent punishment that occurs after a law has been broken? The inflicting of a punishment through the cut, or is it an originary moment in which one sees that there is nothing there: the moment one notices the void itself, and simultaneously realizes that there will never be material large enough, strong enough, to cover over that which has just been encountered, no representative sufficiently robust to stand in its place and make it thematizable? In either case, the loss of a wholeness, of a completion, leads to the desperate attempt at creating some kind of supplement that will fill in that hole, and help to play at making the structure begin to function again as it once did, although such a pure flawless functioning will never be experienced again: returning to that place of a seamless coherent original whole or "all," from this point on, would only ever be a fantasy. So to castrate castration in terms of cutting away from what has already been chopped is not really the best way to describe antherection, its more so that to castrate castration means to bandage the original effect of the infirmity by "banding erect,"[24] propping up a supplemental "prosthesis"[25] to the one that was already there, a banding erect "double";[26] an artificial erection "overlapped" by another, an [overlapping] that "goes over itself again indefinitely," an "unlimited outpouring."[27] Which really means, as Derrida notes, that the original erection itself has been, "in advance, already" an "antherection," which also means that there has already occurred a "castration of castration, an antherection of antherection, and so on to infinity";[28] an original moment of encounter with the void, or the first prosthesis set up to cover over that unbearable encounter, either way, the original, the origin, has been plunged, always already, into an unlimited deluge of pure loss. Fundamentally, it is this experience of pure loss at the level of writing proper that is what is at stake in Derrida's style in general. It is a leaning into, an attempt at true sovereign writing, and in the context of *Glas*, what is gambled upon is the possibility of allowing the explosive force of sovereign writing to offer itself as an alternative, a way out of what for Derrida is the "remain(s) . . . of a Hegel":[29] that speculative dialectics will always be reducible to a fixed absolute moment in which the time of being is the present.

A basic reading of *Glas* that must be stated is that Derrida's fundamental position on Hegel does not change from his early essays on Hegel, in *Writing and Difference* for example (1967), to *Glas* (1974). In "From Restricted to General Economy," Derrida opens his critical position toward Hegel through an engagement with Bataille's thought, a discourse that, for Derrida, compels us to distinguish between "lordship" and "sovereignty."[30] Lordship, for Hegel, gains its position as such through putting its life at stake,[31] but "in order to enjoy what he has won by risking his life" (the prestige of being a

Lord), "the master must stay alive,"[32] a double-edged sword that for Derrida reduces the position of the Lord to a "servile condition of mastery."[33] Such an "economy of life" feels like it is needed because otherwise, "to rush headlong into death pure and simple is thus to risk the absolute loss of meaning,"[34] both in its discursive sense and in a material sense (an abstract negativity). Sovereignty, on the other hand, the position of laughter, of "drunkenness, erotic effusion, sacrificial effusion, poetic effusion, heroic behavior, anger absurdity,"[35] is described by Derrida as being what is "more and less than lordship":[36] it is what opens itself up to "the absolute degree of putting at stake,"[37] which means it is akin to Lordship, but is much more because of this insistence on absolute risk, although (again, in its kinship with Lordship), it too avoids pure abstract negation. "The sovereign operation also needs life," Derrida writes: life sufficient in order for it to "be in relation to itself in the pleasurable consumption of itself."[38] Its solution: to "simulate . . . absolute risk," to stage what looks and feels like a real death, but like the Lord, to skip out on death, just not in the mode of a return to life, not as an ethereal master that lives out its embodied life through the flesh and passions of its bondsman, but through making the world believe that it really did die through a ruse of death, a "simulacrum"[39] of a lifeless corpse, all the while its true self remaining alive behind the scenes, laughing at our shock and horror, its face hidden in the folds of the curtains.

So it is not that meaning is lost "absolutely," as much as it is the absolute of what meaning meant, of what it intended, that is lost forever in the noise of its having fallen into its pluralizing disseminated forms: traces of a trace, speculum angled toward speculum in an endless hall of abstract play, throughout which one hears the laughter behind the stage, "the almost-nothing into which meaning sinks."[40] It is this laughter, the play of the simulacrum that, for Derrida, is "unthinkable for philosophy," is "philosophy's blind spot."[41] There is no such play, no supplement in Hegel. For Hegel, according to Derrida, "nothing must be definitively lost in death," "there must be meaning."[42] Whence the structural tireless work of the *aufhebung*, keeping the entire system safe from true despair, from losing for good what it only wanted to suppress for a moment; the *aufhebung* signifies "a discourse losing its breath as it reappropriates all negativity for itself," in which the gesture of putting something at stake, of exposing it to the negative, would have only been an "investment,"[43] for what was lost was such only for a moment. For Derrida, what Hegel blinded himself to was the implications of a true "radical" negativity, one for which "destruction, suppression, death, and sacrifice" would be "irreversible," unredeemable via an act of sublimation, unresuscitable:[44] a negativity that would sit and remain in the pit. The bedrock of Derrida's critique of Hegel could be summed up as: "always, from the outset, the movement of lost presence already will have set in motion the process of its reappropriation";[45] that is, there is never any surprise in Hegel: what comes to us is what we have already put into the system for safekeeping, there is no supplementarity. Whence the importance of the style of writing Derrida sees in Genet, a writing from the pit, a (dis)style that attempts to truly produce a text of sovereign writing, to have "[known] how to die of laughter when practicing inversion,"[46] and to not "die before having finished writing one's *glas*."[47]

A failed writing, replete with supplementarity, serves another purpose here, probably just as important, that is, its unraveling ruin in the Genet column alongside Hegel, offers itself, for Derrida, as an alternative to Hegelian dialectics, as a way out. He asks the question and then slowly builds up to an answer, explosive in nature: "what would be as it were the absolute knowledge of *glas*?"[48] The beginnings of an answer: "Theft [*vol*] is absolute knowledge."[49] He elaborates: "what was detached . . . was a sign," and "there is where one steals, flies [*vole*]."[50] To thieve, therefore, is to fly away from oneself, to live the melancholia of being ripped away from where one originated, to undergo the experience of a wandering sign, and this is what Derrida sees in Genet: a writer/writing bereft from a home, a home stolen from him, an orphan convict who "keeps nothing close by him, no goods, no works, not himself: no absolute having [*avoir absolu*] as being-close-by-self. He has no place. He slips away and squanders himself,"[51] knowing full well that "one only keeps what one loses. Self."[52]

It is the absence of "chance," of play, of a supplement that Derrida resists in Hegel, whence his being enamored with the opposite in Genet: there, "everything is moved to attach importance to the case of chance," it is a contingency that accompanies the "proper" signifier "[breaking] into language [*langue*], [destroying] itself [there] with an explosion—dynamite—and leaves it as a hole."[53] In Genet the coherency of the subject explodes: "but within myself, this total presence, which is transformed in a bomb of what seems to me terrific power, imparts to the act a gravity, a terminal oneness—a burglary . . . because such a gathering of self [*rassemblement de soi*] cannot take place (not in life)."[54] It is an antiself formation in Genet's writings, a detonation of the autonomous subject, that Derrida sees as the alternative to "the Hegelian gathering close by self."[55]

And so we understand that when Catherine Malabou makes the attempt to reveal how, contrary to Derrida, there is chance in Hegel, serendipity too, that time in Hegel does not just remain fixed in a vulgar state, but becomes infatuated with a necessary future conditional tense (it could happen that way, or it could be otherwise, but something somehow will happen, it's just we won't know what or when, but we will wait for it nevertheless); when she argues that dialectics in Hegel, the *aufhebung* itself, comes to expose itself to its own drama and becomes sublated into a new structure, one of surprise and anticipation, of "seeing (what is) coming,"[56] we understand that when she makes this claim, Derrida's response is incredulous.

For Malabou, this new sense of a subject's relation to the future remains squarely within a logic of dialectics; the subject touches, marks itself, cultivates transformations in itself when it opens itself up to, or anticipates certain "accidents" that are to come, and it is this intimate "relation to its accidents," a relation between its substantiality and its contingent predicates, that for her, comes to "characterize . . . 'plasticity.' "[57] For Malabou, the substance of the subject and the accidents that happen to it "give form to each other" within the framework of this new "anticipatory structure," and they do this in a way that compels us to rethink our relation to futurity.[58] For Malabou, *le voir venir* (to see what is coming) operates as an "interplay within Hegelian philosophy, of teleological necessity and surprise."[59] All of which being a discourse that comes to be interpreted by Derrida as an anticipation, or hope in a future of transformative, creative serendipity that sounds very far away from any intimacy with ruin or death or loss.

Around the heart of his resistance to the possibility of the new form of dialectics that plasticity promises, to the concept of there being two different types of time in Hegel, and to the interpretation of there being a higher sublated structure that heralds a new type of dialectic in the function of *le voir venir*, what moves in that resistance for Derrida is a disbelief, really, that plasticity would have the constitution, the stamina, the stomach, to endure a thing like a bomb, for instance, something that would indeed come to, arrive for, the "opening to sight" of the *voir venir*, but in a "manner [that] would have no essential relation to the future" (xxxiii); that is he does not think that plasticity could handle an event that would arrive, but in a destructive mode so intense that it would end all futures, an *avenir* of a true and pure loss. He writes:

> What would then become of the scheme or of the *hypothesis* inherent to the expression "to see (what is) coming"? What would become of its possibilities . . . [if] [t]o the menace of sight, without even having the time "to see (what is) coming," the plasticity of the plastic exploding in the very hands of all and any 'subject' as a bomb whose timer remains uncontrollable? A bomb which would no longer be regulated or programmed by this "speculative clock" . . . a "speculative subject" [that is] which could not in any case let itself be domesticated . . . re-subjectivized.[60]

Indeed for Derrida, he was never all that interested in a "future" to begin with. He writes to Malabou: "Most of all, what would you do with the fact that I don't even believe of any future for myself and feel at the same time both faithful and radically amnesic?"[61] He remains unconvinced that an "event which we do not see coming, in front of which we do not stand in so far as it does not expose itself, not announce itself in the face of sight or to some sealing of the eye,"[62] he remains doubtful that such an event is even thinkable within the concept of plasticity, and he goes on to wonder really if we would "then have to emancipate what comes (the event), the very concept of the event, from all relation to the future?"[63] Indeed to make oneself a gift of death, the absolute event par-excellence, is what we could not do.

And yet, it has become possible to think plasticity with death, with a force of destruction. In her *Ontology of the Accident*, Malabou, in a way, agrees with Derrida that plasticity is easy to associate with positivity, with changes that take place merely within an easy and hopeful "equilibrium between the receiving and giving of form,"[64] whence a central claim of the text being: "no one thinks spontaneously about a plastic art of deconstruction. Yet deconstruction too is formative . . . [it] has its own sculpting tools,"[65] which is to say, destruction, a pure negating force, might still be thinkable within a kind of dialectical logic in that a pure destructive negation can still be seen as generative, as determining of new forms. The question is will those new forms that emerge from a plastic destructive gesture still be given, held, encountered in a presence, or is it possible to think of forms determined by destruction as being given in the mode of the negative, of nonpresence? Malabou asks toward the end of the book, "Is there a mode of possibility attached exclusively to negation?"[66] That is, not a possibility that would always be reducible to an "affirmation," not a negating gesture that always ends up being sublimated into a new positive form, a thing she writes that "may come into being," but a kind of destroying force that cultivates outside of the mode of bringing into light: "how to think destruction without remission"—without, that is, relevé, resuscitation into the life of presence—"could that," she asks, "be called the *negative possibility*," a "power" of annihilation, "that forms,"[67] but forms new things that would never "come into presence"?[68]

What Malabou is interested in is "a possibility that makes existence impossible,"[69] a full rejection of a future based on hope, or "attached to salvation [or] redemption,"[70] Malabou is interested in thinking an "ego that emerges from unthinkable nothingness, an enigma of a second birth that is not rebirth."[71] It is there, in the sobering reality of a true negative possibility, that "lies the philosophical difficulty that accompanies the thought of an eventness . . . which [owes] nothing, paradoxically, to the thought of . . . any theory of psychic etiology . . . by marking the importance of the brutal and unexpected arrival of catastrophe."[72] And almost in a kind of timed response to Derrida's critique of *le voir venir*, she continues: "the possibility I seek to unearth is precisely the possibility that makes existence impossible . . . the possibility of destructive plasticity, which refuses the promise, belief, symbolic constitution of all resources to come."[73]

What is interesting about the text of *Glas* is that contrary to Derrida's position on the viability of the concept of plasticity, what we find are implications of the very thing he thought plasticity incapable of enduring: a spectral presence of a "negative possibility," the possibility of catastrophe's forming power outside of presence itself can be found in *Glas*. We see this undercutting of Derrida's position on Hegel and plasticity in his own text when the objects that he reads as being endeared to Genet turn against the readings he gives them, and begin to signal other possible destinations that Derrida certainly could not have anticipated, least of all destinations that would end up undoing the very heart of his critique of Hegel.

I have mentioned the significance of flowers in Genet, but not the Vaseline tube and spit. Certainly, like the former two objects, these things as well function in Genet in ways that confirm and support a logic of deconstruction in that they can all be read as affirmative of how the self never escapes an inevitable dissolution of its proper and coherent formation (and Derrida's point has been that Genet is such an interesting writer for the ways in which the subject's loss of itself is pleasurable to it not frightening). For example, Derrida finds in Genet that the words like "saliva," "sperm . . . glair, curdled drool, tears of milk, gel of vomit—all these heavy and white substances are going to glide into one another . . . stretched out (on)to the edge of all the figures and pass through all the canals . . . [their] description is caught in a veil."[74] A little further on, "the agglomerate web-veil-spittle" is seen to "[take] on other contents," its "reference" (or referent) "can always . . . be turned inside out like a glove. . . . The suspension of the veil or of the spit, the elaboration time of the excrement is then *also* the indecision between reference's two directions [*sens*]."[75] The interest spit holds for Derrida is the way it rubs against other strange signifiers in ways that keep the indeterminacy of the veil charged, that keep the position of the referent structurally open, available for substitution. What is not pressed as much, however, are the sexual valences that these signifiers hold.

A similar gesture of overlooking the sexual undercurrents of Genet's style characterizes Derrida's reading of the Vaseline tube. He cites Genet:

> I was dismayed when, one evening, while searching me after a raid . . . the astonished detective took from my pocket, among other things, a tube of Vaseline . . . one whose ends was partially rolled up. Which amounts to saying that it had been put to use. Amidst the elegant objects taken from the pockets of the men who had been picked up in the raid, it was the very sign of abjection, of that which is concealed with the greatest of care. . . . But lo and behold! That dirty, wretched object whose purpose [*destination*] seemed to the world . . . utterly vile, became extremely precious to me.[76]

Again, in Derrida's reading of this passage, it is not the sexual nature of the object or its context that guides him, but simply Genet's counterintuitive (deconstructive) adoration for that which represents pure abjection to the rest of the world. Again, what Derrida continues to draw from Genet is how this object "illuminates" an intoxicating "[fall]" to the tomb.[77] What Derrida has left unsaid is that all of these objects—flowers, a tube of Vaseline, saliva—are queer objects, and not just queer in a vague sense that inspires a double take, a second brisk reading, but queer in a negative sense: more precisely, all of the objects are over invested with (hyper) sexual charge, and are figures that are upheld by a queer negative force. Such a force can be easily read and appropriated by a deconstructive logic as being likeminded, as harboring the same agenda (in relation to structure and points of origin), but these objects, antisocial all of them (in how they have been put to use), also locate, decidedly, *Glas* in a queer camp of negativity and, if Lee Edelman is to be believed, operate at the base level of the destructive drive, and move and affect from that (non)place the symbolic order.

Edelman, in his *No Future: Queer Theory and the Death Drive*, is completely invested in severing the notion of queerness from identity; he is not interested in repairing the symbolic so that it can accommodate new figures within its structured order of intelligibility. What interests him is destruction of that order once and for all: "Are we willing," he asks, "to accept that the figural burden of queerness . . . is that of the force that shatters the fantasy of imaginary unity, the force that insists on the void . . . always already lodged within, though barred from symbolization: the gap or wound of the Real that inhabits the Symbolic's very core?"[78] Lacan is central for Edelman in his theorizing of queer psychic life, in that queerness for him "[embodies] the remainder of the Real internal to the Symbolic order,"[79] a level of remains at which also operates, surges and erupts jouissance, or enjoyment. It is the concept of enjoyment therefore, a queer jouissance, that guides Edelman in his thinking of queer as what "dissolves . . . the consistency of a social reality that relies . . . on the structures of Symbolic law."[80] It is at the subterranean level of the death drive that Edelman locates queer, a destructive force that, for him, "always insists as the void in and of the subject, beyond its fantasy of self-realization, beyond the pleasure principle."[81]

If anything describes the nature and significance of Genet's objects, certainly it would be this. Three pages after reading the scene of the detective finding the Vaseline tube, a scene involving spit in Genet's *Miracle of the Rose* makes it impossible to overlook the inextricable link between queer sexuality and death. I quote from the Genet text itself, the same scene cited in *Glas*:

> As soon as I saw him approaching, I realized that my time had come. They were going to proceed to my execution. . . . When I arrived the seven big shots . . . stopped talking. Van Roy cried out in a joyous tone: "Here we go, boys! Fifty feet away!" He placed himself in front of me, at the said distance,

> and yelled: "Open your mug," . . . I opened my mouth. . . . He took his distance again, leaned over a bit to the right, aimed, and spat into my mouth. An almost unconscious moment of deglutition made me swallow the gob. The seven of them howled with joy. . . . Still shaking with laughter, Deloffre spat into my eyes. The seven of them took their turn, in fact several times, including Divers. I received the spit in my distended mouth, which fatigue failed to close. . . . I was hit in the face and was soon slimier than a prickhead under the discharge. I was then invested with a deep gravity . . . a trifle would have sufficed for [it] to be transformed into a courtly one and for [him] to be covered not with spit but with roses . . . for . . . the gestures were the same.[82]

It is very easy here, for example, to see how the drive operates in oscillation between destruction and queer sexual desire. For Edelman, "queerness embodies this death drive . . . by figuring sexuality's implication in the senseless pulsations of that drive. Deidealizing the metaphorics of meaning on which heteroproduction takes its stand, queerness exposes sexuality's inevitable coloration by the drive."[83] It's interesting, when Lacan first begins to teach on the drives in his seminar VII, he opens onto the concept via Sade, and what he loved about the potentialities of a destructive force within the social. He quotes from Sade: "Without destruction the earth would receive no nourishment and, as a result, there would be no possibility for man to reproduce his species . . . the vices . . . of our social system . . . are creative . . . nature wants annihilation; it is beyond our capacity to achieve the scale of destruction it desires."[84] We see here in Lacan, through Sade, a quiet implication of a productive determining agency as being embedded in the death drive.

Malabou does note the relevance of thinking the death drive as a possible example of a destructive force that still is generative: "Freud knew that the death drive creates forms," yet, she goes on to ask, "how does one render the death drive *visible*?"[85] Later in *Ontology of the Accident*, she abandons the death drive as a viable conduit for a negative possibility; for Malabou, when Freud in his later work develops the concept of there being two different drive natures, one of *eros*, the other of death, of "negation," he associates negation with "the instinct of destruction," destruction in this context being abstract annihilation, and for Malabou, this does not describe the phenomenality of a negative possibility: what Freud is doing here, for her, is confusing the negative possibility with "the destruction drive," that is to say negativity here "becomes annihilation,"[86] it would not lead to the generating of a new form outside of presence.

But what Edelman is doing when he thinks the relation of a destructive queerness and the drive, is keeping his notion of the drive's effects from any easy reduction to an abstract annihilation. In often uncanny moments of what can only be described as anticipation (or mimicry) of Malabou's destructive plastic thesis, Edelman often asks of queer negation its capacity to produce new forms outside the mode of presence. His dialogue with Berlant, *Sex, or the Unbearable*, is another text created out of a need to stare the negativity of sex, of queer sexuality more specifically, dead in the face, a theoretically creative yet critical defense of the antisocial position within queer theory.[87] There, when thinking about the negativity attached to the movements and noise of queer, for Edelman, it's much more important to ask what does it mean to be unable to think sex without continuing to "come back to the unbearable in the encounter with negativity" (one thinks of the uses of spit in Genet).[88] Negativity for Edelman "is unchanging as structure because negativity structures change."[89] The question remains however, "what sort of change are we talking about"?[90]

I would like to step out on a limb and suggest that the change that would interest Edelman most would be one of a negative possibility, a hit by a queer negating force of a true catastrophe (onto or within a violent structure of regulating and suffocating normativity), but one that still forms, just in ways that are not recoverable in or by presence, but that remain, giving themselves, effectively and affecting, in a negative mode. Toward the end of their text, Edelman reflects on the relation between surprise and negativity: "It takes a lot of attention and a willingness to be seized by something one cannot know in advance, to retain the capacity for surprise before what offers itself as what is; it takes a continuous resistance to one's englobement by 'the world.'"[91] What "offers itself as what is," I think in this context, is a destructive yet generative agency that is indeed anticipated, but when it arrives always surprises, not least of all in the mode of queer psychic life. To encounter a queer negativity for Edelman is to open oneself up to "an otherness that undoes our image of the self."[92] The trick in his formation here is to be undone is to also undergo a change, a negative possibility that leads to a transformation. For Edelman, once we encounter a queer negating force, "if that negativity entails destruction, such destruction enables change."[93] He continues on to describe this change in what could only be described as plastic terms: a queer destruction enables change, "not . . . by redeeming negativity [not by relieving it in a surfacing moment into presence] but precisely by *enacting* it, by imposing it as the condition of being open to (ex)changes [ek-stasis] that promise us nothing more than being bound to them, forever."[94]

It turns out that Derrida does not escape Hegel through Genet. The very thing that he saw in Genet as a critique and alternative to a Hegelian dialectical metaphysics of presence—a deconstructive style of writing in Genet, mostly written through the voices and affects and bodies of queer subjects—turns out to be caught up within another form of dialectics, a destructive plasticity. Which is to say that the writing that Derrida admires so much in Genet for the way in which it leaves Hegelian dialectics behind, a writing marked by a queer destructive force, remains attached to dialectics, just in a plastic form:

a new mode of dialectics in which a queer negativity transforms the subject, leaving it changed, but does so in ways irreducible to presence. That is, in Genet we see a queer negating force operating as a negative plastic possibility that forms the subject out of an "unthinkable nothingness."[95]

NOTES

1. "Real" understood in the Lacanian sense of that which resists thematization.
2. Jacques Derrida, *Glas*, trans. John P. Leavey (London: University of Nebraska Press, 1986), 1.
3. Lee Edelman, *No Future: Queer Theory and the Death Drive* (Durham, NC: Duke University Press, 2004), 6–7.
4. Ibid., 7.
5. Ibid.
6. Lee Edelman and Lauren Berlant, *Sex or the Unbearable* (Durham, NC: Duke University Press, 2014), 63.
7. Ibid., 65.
8. Jacques Derrida, "A Time for Farewells: Heidegger (read by) Hegel (read by) Malabou," preface to Catherine Malabou's *The Future of Hegel: Plasticity, Temporality and Dialectic* (London: Routledge, 2005), xlvi.
9. Derrida, *Glas*, 262.
10. Derrida, *Glas*, 1.
11. Ibid.
12. Ibid., 2.
13. Ibid., 25.
14. Ibid., 2.
15. Ibid.
16. Ibid., 32.
17. Ibid.
18. Ibid., 47.
19. Ibid., 17.
20. Ibid., 131.
21. Ibid., 130.
22. Ibid., 131.
23. Ibid.
24. Ibid., 138.
25. Ibid.
26. Ibid., 201.
27. Ibid., 130.
28. Ibid.
29. Ibid., 1.
30. Jacques Derrida, "From Restricted to General Economy: A Hegelianism without Reserve," in *Writing and Difference* (Chicago: University of Chicago Press, 1978), 254.

31. Ibid., 254.
32. Ibid.
33. Ibid., 255.
34. Ibid.
35. Ibid., 256.
36. Ibid.
37. Ibid.
38. Ibid.
39. Ibid.
40. Ibid.
41. Ibid., 257.
42. Ibid., 256–57.
43. Ibid., 257.
44. Ibid., 259.
45. Jacques Derrida, "The Pit and the Pyramid: Introduction to Hegel's Semiology," in *In Margins of Philosophy* (Chicago: University of Chicago Press, 1982), 72.
46. Derrida, *Glas*, 129.
47. Ibid., 61.
48. Ibid., 225.
49. Ibid., 173.
50. Ibid., 167.
51. Ibid., 206.
52. Ibid., 207.
53. Ibid., 236.
54. Ibid., 174.
55. Ibid.
56. Catherine Malabou, *The Future of Hegel: Plasticity, Temporality and Dialectic* (London: Routledge, 2005), 17.
57. Ibid., 11.
58. Ibid.
59. Ibid., 13.
60. Ibid. Derrida's preface to *Future of Hegel*, xxxiv.
61. Ibid., xxxiii.
62. Ibid.
63. Ibid.
64. Catherine Malabou, *Ontology of the Accident: An Essay on Destructive Plasticity* (Cambridge: Polity Press, 2012), 3.
65. Ibid., 4.
66. Ibid., 73.
67. Ibid., 75.
68. Ibid., 85.
69. Ibid., 88.
70. Ibid., 89.
71. Ibid., 90.
72. Ibid.

73. Ibid., 88.
74. Derrida, *Glas*, 139–40.
75. Ibid., 141.
76. Ibid., 142–43.
77. Ibid., 144.
78. Edelman, *No Future*, 22.
79. Ibid., 25.
80. Ibid.
81. Ibid.
82. Jean Genet, *Miracle of the Rose*, trans. Bernard Fretchman (New York: Grove Press, 1966), 266–68.
83. Edelman, *No Future*, 27.
84. Sade quoted in Jacques Lacan's *The Ethics of Psychoanalysis: The Seminar of Jacques Lacan: Book VII* (London: Routledge, 2008), 259–60.
85. Malabou, *Ontology of the Accident*, 18.
86. Ibid., 84.
87. Edelman and Berlant, *Sex, or the Unbearable*, 111. In a characteristic stroke of gorgeous style, Berlant sketches what such a position feels like at times (although part of the project of this text has been to not just jettison this critical perspective in queer theory, but to think twice about it): "ways its unbearable to be partnered by one's brokenness and saturated by the bad, wearing world."
88. Ibid., 121.
89. Ibid.
90. Ibid.
91. Ibid., 120.
92. Ibid., 33.
93. Ibid.
94. Ibid., 33–34.
95. Malabou, *Ontology of the Accident*, 90.

BIBLIOGRAPHY

Derrida, Jacques. "A Time for Farewells: Heidegger (read by) Hegel (read by) Malabou." In *The Future of Hegel: Plasticity, Temporality and Dialectic*, vii–xlvii. Translated by Lisbeth During. London: Routledge, 2005.

———. *Glas*. Translated by John P. Leavey. London: University of Nebraska Press, 1986.

———. "From Restricted to General Economy: A Hegelianism without Reserve." In *Writing and Difference*. Chicago: University of Chicago Press, 1978.

———. "The Pit and the Pyramid: Introduction to Hegel's Semiology." In *Margins of Philosophy*. Chicago: University of Chicago Press, 1982.

Edelman, Lee. *No Future: Queer Theory and the Death Drive*. Durham, NC: Duke University Press, 2004.

Edelman, Lee, and Lauren Berlant. *Sex or the Unbearable*. Durham, NC: Duke University Press, 2014.

Genet, Jean. *Miracle of the Rose*. Translated by Bernard Fretchman. New York: Grove Press, 1966.

Lacan, Jacques. *The Ethics of Psychoanalysis: The Seminar of Jacques Lacan: Book VII*. London: Routledge, 2008.

Malabou, Catherine. *The Future of Hegel: Plasticity, Temporality and Dialectic*. Translated by Lisbeth During. London: Routledge, 2005.

———. *Ontology of the Accident: An Essay on Destructive Plasticity*. Translated by Carolyn Shread. Cambridge: Polity Press, 2012.

Chapter 12

Plasticity of the Mind

Reflecting on and Discussing Marcus Aurelieus's Meditations *with Catherine Malabou*

Georgia Mouroutsou

A CRISIS, THE EXPECTATIONS FOR THE PHILOSOPHERS AND HOW CATHERINE MALABOU MEETS THEM IN *WHAT SHOULD WE DO WITH OUR BRAIN?*

A long time ago, philosophy was a princess, her kingdom reigning over all inquiry. The philosopher, in Plato's narrative, had to fulfill the extraordinarily demanding expectation of being an expert in all fields. And as philosophy came down to earth with Aristotle, its reign was divided into three parts, followed by the philosophy of action and the philosophy of production: a kingdom divided, and less divine, but a kingdom nonetheless. The Stoics followed with a concept of philosophy as divided into three parts, which, at the same time, are organically bound like the limbs of an organism. Against this holistic framework, the later Stoics practiced philosophy as an art of living and, in particular, as therapy of the soul, pursuing and transforming the Socratic model of soul care.

These times of glory and greatness are long gone and have given way to further developments and intellectual achievements throughout the ages. After many transformations of philosophy's meaning and role, here we stand in our age of wisdom, expertise and rich and diverse ways of doing philosophy, all of which are accompanied by the isolation and fragmentation of disciplines and forces in and beyond philosophy, and, to make matters worse, by ungrounded ideological hostilities. As I see it, the role of the philosophers in this age of ours is to learn how to effectively lead the way again—not for the sake of leading, but for the sake of solving the new problems we are confronted with. We need to get out of our comfort zone, conduct genuine dialogue, and apply new theoretical reconstructions and frameworks to the many crises we encounter: to name one of the most serious ones—the illness of depression, which, as affecting greater and greater numbers of people,[1] has been characterized as *the* disease of our age: the age of unimpeded and unquestioned utilitarianism.

It is thereby necessary to put aside our sympathies or, even worse, ideologies, and bring together different and even opposite directions of thought and science in order to construct theoretical frameworks and provide practical solutions. The responsibility for genuine dialogue against the above framework and tradition, of which only a tiny bit I indicated in broad strokes, lies on no one else but the philosopher, if I might raise high expectations for her, with a Platonic touch in these expectations. The thought of Catherine Malabou offers a unique model in the above respects, unique, as all models of excellence are, in accomplishing a fruitful dialogue between philosophy, neuroscience, and psychoanalysis.

In her *What Should We Do with Our Brain?* Catherine Malabou analyzes the brain as a work, a work in progress in which we participate,[2] and characterizes our time as one of neuronal liberation, a liberation that is not to be taken as a given, but rather as a potential that we have yet to fulfill, and an ideal to live up to. The rich semantic range of the word *plastique* and *plasticité* offers the starting point for Malabou's investigation. The meanings of the word "plasticity" in French are both passive and active, and range from the one following extreme to the other: from the capacity to be molded to the ability to shape and transform predestined constellations, and, even, on the other end, exploding.[3]

This analysis is evidently not merely semantic and theoretical, but strongly protreptic. Malabou, namely, invites us to actualize the above potential and liberate ourselves; the transition that is to take place being the one from the omnipotent demands for passive adaptability and flexibility to resisting those demands and unleashing new ways of living—creative and resistant, even explosive, plasticity.

Instead of asking what the essence of the brain is, which is inappropriate according to Malabou, because the brain is a work in progress, she asks what we should do with our brain. I would like to make three points of relevance: first, to ask this question presupposes that we *can* do something with our brain, since the brain is not fully determined, but equipped with the power to transform itself; second: Malabou's undertaking is not motivated by a desire to see quantifiable results; that said, she invites us all, including utilitarians, to play a game that sounds familiar: what should we do with our brain so that we get the greatest benefits and the least costs for the greatest number of people? The game is transformed, though. For Malabou's question invites us to do things that are not measurable with the aid of utilitarian scales: to wonder at and learn from the brain as a model. Third: by so doing, we entertain an image of the world to come, and might even construct ideals for the future in the very present moment, instead of unsuccessfully trying to pursue and keep pace with the current trends, and thereby skipping the present moment itself.[4]

When Malabou turns her attention to the hotly debated issue of reductionism à la "you are your synapses," she puts forward a very interesting theoretical option: she does "in no way presume to contest the hypothesis of neuronal and mental continuity or to play the game of antireductionism,"[5] the former presupposed by most neuroscientists,[6] the latter represented by many philosophers, who fall into this trap as if no other stand were possible.[7] For, Malabou is a materialist as most neuroscientists, but, in contrast to them, she reflects on the gaps and leaps, which are involved in the transition of the neuronal to the mental, a transition whose continuity she does not take for granted. She, instead, escapes the strict alternative and opens up a third path between reductionism and antireductionism on the basis of her expertise in Hegelian dialectics. The path from the neuronal patterns to the mental images is characterized by struggles between the biological and mental against the other, and "thought is nothing but nature, but a negated nature, marked off by its own difference to itself."[8]

INVITING MARCUS AURELIUS AS AN ALLY INTO THE DIALOGUE WITH CATHERINE MALABOU

Inspired by Malabou's challenging analysis of the term "plasticity," I will invite Marcus Aurelius in what follows, the Roman emperor and Stoic philosopher of the second century CE, to corroborate and enhance the French philosopher's undertaking. I will focus on Marcus Aurelius and the nature of mind as he describes it in his *Meditations*, because he is a materialist, like Malabou. By focusing on a materialist, we have, furthermore, better chances when engaging with the neuroscientists. In addition, researching topics in Marcus like the plastic power of the mind and, moreover, the dimension he gives to acceptance to mention only two themes[9] can help us with the cure of affective disorders, and, in particular, depression, which, as I said in the introduction, has become *the* illness of our era.[10]

To start with, I will take a stand with regard to the opposition between reductionism and antireductionism, and thereby avoid wishing to provide a knock down argument for any party. Nor will I choose the path that Malabou chose to take, but propose a fourth way to deal with the dilemma, and thereby pave the way for Marcus Aurelius. Afterward, I will turn to the special power that the mind has, according to Marcus Aurelius, against the background of his materialism, namely, the power to shape and act upon what it is confronted with, a power that distinguishes it, though it is corporeal itself, from other bodily entities. Finally, I will focus on the power we have, based on the mind's distinctive power, not just to passively accept anything devastating, but to transform it and, as such, that is as transformed, to love it. This discussion is highly relevant for and conducive to the debated issue of acceptance in cognitive behavioral circles, which I draw upon because cognitive behavioral therapy has been proven to be the most efficient in dealing with depression, and is, therefore, the psychotherapy that nurtures our hope of further enhancement of and contribution to the cure of affective disorders.[11]

After highlighting the active power of the mind to shape and act, I will end with a suggestion, by turning to something forgotten, in the history of ancient philosophy, at least, after Plato: to act is not the only power; to be acted upon is a power as well. This will be my critical note to Malabou's overstressing the active power to change and transform, and her not considering how powerful being acted upon might be in itself and how an underestimated but essential part of plasticity it amounts to. My critique sharpens even more the critical spirit of Malabou's undertaking along her own lines.

LOGOTHERAPY MATTERS: SUSPENDING THE DEBATE BETWEEN REDUCTIONISM AND ANTIREDUCTIONISM

To put my cards on the table: for my part, and contrary to Malabou, I play the game of nonreductionism because it is a very serious game to play to the bitter end, I think, when the life of others is at stake, as is the case when it comes to clinical depression. Were the mental events completely reducible to physical events, the latter would be sufficient to necessitate moods in us, but, on the contrary, not only the chemical activity in our brain but also our judgments *qua* judgments have immense power and exert considerable influence on our moods. On these grounds, medication is in certain cases and stages of affective disorders necessary, but, merely on this basis, long-term cure cannot be guaranteed. The avoidance of recurrence of and relapses into depressive episodes can be enhanced by logotherapy that focuses, beside behavioral intervention, on the intervention into or even change of false thought processes. The latter are well rooted in deeper, negative belief-schemata, a system of unfavorable beliefs which is activated when triggered by a life event. And stronger evidence has been added: a recent study has shown that cognitive behavioral therapy alone and without any addition of medication can cure severe depressive states, which has corrected a previous myth that the use of medication is the best and only treatment in the case of severe depressive episodes.[12]

The way that thoughts influence moods provides evidence for the lack of continuous and automatic transitions from the physical to the mental, that is for nonreductionism and for the power we have over what we may have falsely thought to overcome us: so-called feelings, emotions, or moods.[13] Admittedly, the partisans of reductionism will put forward a battery of arguments and maintain either that any influence on attitudes and thought chains is grounded in a previous influence on brain chemistry, or, in a less fine-grained manner, that future medication might prove logotherapy obsolete.

Though I just put my cards on the table and chose the game to play, it is not my intention here to analyze the dialogue or the parallel monologues that take place between reductionists and antireductionists, and even less to offer definite solutions to the intricate problems involved. Because logotherapy works here and now, and because this is what matters, when we focus on the cure of affective disorders and the enhancement of patients' lives, I do not even mind to suspend the judgment and decision about the debate, and focus, instead, on what we should not postpone elaborating on: the enhancement of logotherapy itself for the sake of the cure of affective disorders. For, we cannot wait for the solution of the debate in order to do so. Exactly here Marcus Aurelius enters the picture as someone not interested in participating in theoretical debates, but in formulating precepts and exercises about the art of living, and whose self-reflections can contribute to further improvement of cognitive behavioral therapy. By taking advantage of the special power that is allotted to our mind, Marcus Aurelius has been attempting to cure his passions and attain moral progress. It is on the basis of the power of judgment and thought that cognitive behavioral therapy has been introduced,[14] and it is on this basis and against the background of later Stoics like Epictetus and Marcus Aurelius that such a therapeutic strategy cures depressed patients to the point that they might not even need medication again.

MARCUS AURELIUS ON THE MIND'S PLASTIC POWER AGAINST THE BACKGROUND OF HIS MATERIALISM

As all Stoics, Marcus limited reality to bodies: anything real—including god, soul, or intellect—is a peculiarly qualified body. Only bodies have the power to act or to be acted upon, or both, whereas anything deprived of corporeality is deprived of causal efficacy and is, therefore, doomed to be inert:[15] such as space and time, among other incorporeals, which at best provide the background for the interaction of bodies. To move from bodies to the deeper level of their principles: in the world of the Stoics, the slightest physical change—and all change is physical according to their view—manifests the interaction of two principles, active god and passive matter, which are the only permanent items in their universe. Despite their radical corporealism, the Stoics do not attempt to reduce nonmaterial to material properties. For, body does not only possess the material qualities of solidity, bulk, and weight, but is also alive and intelligent throughout, its intelligibility and materiality being separable only in thought.[16]

Against this background and tradition, Marcus reminds himself of his mind's power in many passages of his *Meditations*; for instance, in the following reflection in the fourth book, in which he refers to mind as the ruler within us, and in which he highlights the easiness with which it transforms and adapts itself to whatever might be happening in our life:

> When the ruling power within us is in accordance with nature, it takes a stand toward what happens so that it always adapts itself easily to what is possible and what is presented to it. For, it does not like any specific material, but sets out for its preferred item with reservation, and converts what emerges against expectation into material for itself, like fire when it prevails over the things that fall upon it, by which a little lamp would have been extinguished. A blazing fire, instead, appropriates the things that are piled upon it in the quickest way, and devours them, and raises higher with their aid. (Part of 4.1)[17]

The mind's power is wondrous (part of 5.20): "For the mind adapts and converts everything that impedes its activities into something that advances its purpose, and a hindrance to its action becomes an aid, and an obstacle on its path helps it on its way."

And the wonder continues. Along the same lines, and even more succinctly, using the terminology of the Early Stoics for the intellect, which he calls the ruling center, Marcus notes to himself (6.8): "The ruling center is that which arouses itself, adapts itself, makes itself what it wants to be, and makes everything that happens appear to itself as it wants."

Though grounded on a theoretical edifice,[18] philosophy, as Marcus Aurelius describes it and practices it in his *Meditations*, is not merely an intellectual undertaking, but an art of living.[19] The goal when exercising this craft is to live according to nature by making moral progress and becoming a better person.[20] In this craft, as in every craft, both form and matter are involved: for, all craftsmen turn a material, the subject matter of their craft, into an ordered product.[21] It is with the aid of intellect that we form the material offered to us, namely, whatever happens in our present moment. This material is shaped by the judgments we formulate about and the evaluations we make of the current events. The power of judgment is the power with which our intellect is endowed.

The active power that our mind can exercise on the basis of judgment is enormous.[22] For, what shapes our life are not the happenings themselves, but the way we decide to deal with and use them. Any happenings that come our way, instead, take on the passive role of matter. They provide nothing more—and nothing less—than the necessary material that will be shaped by our mind, though they might initially strike us as mighty, positively or negatively powerful in themselves, and as capable of shaping our life, by way of enhancement or debilitation, respectively. Even what seems to be detrimental to a flourishing life, like sickness, poverty, or social exclusion[23] can be interpreted, and should be interpreted according to Marcus Aurelius, as a challenge to be met for the sake of moral progress. Our mind is able to turn a disadvantageous situation into one conducive to our becoming better and more praiseworthy with regard to virtue than before the devastating event took place if and only if we decide to do so. When we fixate our gaze on the possession and preservation of a particular preferred item, like the opposites of the above mentioned, for example, health, wealth, or fame, our mind does not act in accordance to its own specific nature and power. We should thereby take on full responsibility for deciding to delimit our mind's power and restrict its potential.[24]

Marcus Aurelius makes clear throughout his reflections that it is up to us to use our intellect according to or against its own nature. In the tenth book, we furthermore read that the accompanying affectivity when working on all possible current material is that of enjoyment, and Marcus Aurelius adds some detail about the contrast between the intellect's motion and other corporeal entities' motions; in contrast to the latter, the former cannot be impeded:

> What can be done or said in the soundest manner, given this matter? For whatever that may be, it is possible for you to do or say it, and you should not make the excuse that you are being hindered. [. . .] For you should regard as enjoyment all that is possible for you to do according to your specific nature; and everywhere this is possible for you. It is not granted to a cylinder to act out its specific motion everywhere, nor to water, nor to fire, nor to anything else that is governed by nature or irrational soul; for there are many things that prevent them and stand in their way. But mind and reason can advance as they are by nature, and as they want, through everything that is adverse. Set before your eyes this easiness with which reason can move through everything (as fire moves upwards, a stone downwards, a cylinder down an incline) and seek after nothing more; For, all other impediments are either of the poor dead body, or do not break nor do any kind of harm without judgment or without reason itself giving in. Otherwise the person affected would at once be bad. For, in the case of all other objects: whichever bad thing may fall upon any of them, what is affected by that becomes worse; whereas here, if one must say, the human being becomes even better and more praiseworthy when properly using what falls upon him. (Parts of 10.33)

We should always keep in mind that, according to Marcus Aurelius, the intellect is both active, in the above sense, and passive. For, at the same time when transforming the material it receives into its own, it is being transformed.

Intellect is corporeal,[25] and, as such, is necessarily affected and moved by whatever it encounters,[26] when appropriating what is presented to it at each and every moment. According to Marcus Aurelius's analysis, our intellect is, therefore, plastic:[27] it is both capable of being shaped and shaping; to put it in Stoic terminology, the intellect is capable of both being acted upon and acting. Thanks to this constant plastic transformation, our life is a work in progress; interestingly enough, and in contrast to all production, including the artistic one, Marcus Aurelius characterizes this work in progress as complete in each and every moment.[28]

MARCUS AURELIUS ON OUR POWER TO BE DELIGHTED WITH AND LOVE WHATEVER HAPPENS TO US

In *Med.* 10.1, Marcus wonderfully formulates what he is conducting in the entire *Meditations*: an inner dialogue with his soul while he is morally progressing from his occurrent to ideal self:

> My soul, will you ever be good and simple and one and naked, and more evident than the body which envelops you? Will you ever taste what a loving and affectionate disposition is? Will you ever be complete and lacking nothing, yearning after nothing, and desiring nothing [of the pleasures of the transient sensible things, or fame that will sooner or later turn into dust, GM], be it animate or inanimate, for the sake of enjoying your pleasures? And (will you ever) not wish for more time, in which you can have enjoyment for longer, or a well-situated place or country or seasonable climate or more suitable people? Will you be, instead, satisfied, with your present circumstances and delighted with everything that is present and will you convince yourself, my soul, that everything you have in the present has come from the gods, and everything is well for you and will be well, which is pleasing to them, and which they will give you in order to preserve the perfect living being,[29] which is good and just and beautiful, and which generates and holds together and encompasses all things, and embraces them when they are dissolved so that others of like nature come into being? (Part of 10.1)[30]

How can Marcus become so passionate here, Marcus who, as a Stoic philosopher, successfully attempts to disattach himself from emotional involvement throughout the *Meditations*? How can he speak—not of enjoyment, as in a previously quoted passage, but of the emotional crescendo of delight, when it comes to present happenings, he who has been aiming at curing himself from all affective elations? The answer we can give on a basic level is that he did not wish to free himself from all affective emotional states, after all, but to refrain from one particular type of passions someone might succumb to when jumping from one fleeting pleasure to the next due to attaining and briefly preserving a preferred indifferent item, and to break free of the vicious circle of the desire-satisfaction and desire-frustration, which arises when one is caught up in the whirl of passions related to fleeting indifferents.

It was easy to draw this conclusion with the aid of the Stoic distinction between good passions and passions that are to be eradicated,[31] but there's the rub: what does it mean to *love* whatever one encounters in the present moment? Formulating the statement that one should love the present happenings goes much further than highlighting the active and transformative power of the intellect, though, admittedly the affective attitude of love is grounded in mind's nature. Let us think of the following example: one is listening to the doctor's diagnosis of schizophrenia in the very present moment. What does it mean for her "to love what is happening in the present moment"? Should she jump to the ceiling and make explicit how she adores what is happening in this very second? This cannot be the meaning of loving the present happenings according to Marcus. His Stoicism is far from being so unreasonable.

In order to explain what Marcus Aurelius means, instead, I will briefly address two relevant and interconnected issues here at stake: the plasticity of mind and its affective counterpart, that is, love, on the one hand, and plasticity of temporality, with the particular focus on the present moment, on the other. To the first aspect: Marcus Aurelius speaks of accepting, loving, and embracing whatever happens,[32] but does not thereby prompt himself to passively swallow anything as a bitter pill or to merely acknowledge the unfortunate event as such, namely, as a fact, in our example the fact of the particular diagnosis on the particular day of the calendar. For, we do not live our lives by filling our suitcases with facts. Instead, we fill our lives' suitcases with interpretations, interpretations which will be constantly changing as we constantly renew ourselves through the passage of time and through our passage in time, if we wish to follow Marcus's advice to himself to be constantly renewing himself.[33] Our mind is free to find a best possible interpretation of the present moment that will then become past so that we become even better than before: before the "bad luck" struck. One cannot "love" the diagnosis of schizophrenia, of cancer, or, anything suchlike, in themselves, but one can love what one makes out of them, for we are the artists who mold the diagnoses and not vice versa.[34]

To the second aspect of plasticity of time, and, in particular, the present time: one cannot "love one's fate," as Marcus advises himself to do in the very present moment. The most and best one can do then, in our example of diagnosis of schizophrenia, is to perceive what is happening without adding any value judgment.[35] To become able to love such a happening presupposes to creatively reshape and reinterpret what has happened, which in the case of such devastating incidents takes a lot of time and an equally lot of suffering.

This reshaping can only happen against the background of one's entire life, and definitely after the present moment has become past—remote past rather than recent—and, more importantly, after one has integrated it into the context of one's an entire life, and has been able to see why certain aspects have contributed to one's developing sensitivities and capacities that one had not imagined one had beforehand. Therefore, this kind of love is not a love at first sight—or, rather struck—but can be developed only in the long run. If one makes it, one will be able to lovingly embrace one's entire present and so experience the present as embracing his whole life, past and future and not as a point in time.[36] To sum up, "to love whatever happens in the present moment" goes beyond the very second in which the diagnosis in our above example is uttered and demands the transformation of a whole life's narrative, an undertaking which Marcus Aurelius is conducting in the entire *Meditations*.

A LAST SUGGESTION

I will end my reflections with a suggestion that critically corroborates Malabou's critique of capitalism. She highlights the power to produce, to make and shape as part of the meaning of plasticity, and decides against the mere capacity, because capacities may be left unactualized, after all,[37] whereas the passive counterpart—to be shaped—she deprecates as flexibility. I suggest, instead, that we start reappreciating passivity as a power as well. In a society which obsessively highlights and advertises activity, action and production, whereas affectivity and passivity of any sort, which might cause a break from the so-called productive modes, is looked down upon, philosophers take on the responsibility to help us all pause and evaluate the passive power as it deserves. This sensitive halt will help us attain some healthy critical distance from what is generally promoted. Furthermore, we will thereby become able to reevaluate the power to be affected and acted upon in people who have the tendency to, and experience of, clinical depression, instead of considering them as unproductive and costly. By doing so, we will grasp the opportunity to detect in them what makes us all human, instead of cultivating the attitude according to which we have to help the disabled to comply with what our society promotes and what we follow without any critical reflection.[38]

In this spirit: shall we not decide to appreciate the power of receiving and being shaped—in genuine dialogue—as an essential part of what makes an individual and a culture "plastic"?[39] Is it not also the power of being shaped and affected, the power of giving space and making room and listening to the other as other (whatever and whoever this other is), that produces great students, great teachers, and, in general, great individuals and cultures, even if this greatness is fragile and vulnerable, as all greatness in individuals and cultures is?

Shall we not listen to other, "disaffiliated" citizens of our world, as Malabou labels them, and their power of being affected, citizens who can teach us a bit of their accumulated and transformed wisdom after having been cured? Or, shall we, instead, with Aristotle, confine happiness to the activity of doing, and exclude ourselves from the entire range of the meanings of plasticity, which includes being shaped besides shaping? It is up to us to decide which way to go.

NOTES

I am indebted to Antonio Calcagno, Paulo Ferreira, Stephen Lofts, and David Dozois for helpful comments and stimulating discussions.

1. Because of its prevalence, depression has also been characterized as the "common cold" of mental illness. For, 10 percent of the population suffers from depression, and, according to recent estimates, 17 percent of the population has one depressive episode over the course of their lifetime. In contrast to common colds, the impact on the individuals is drastic on all possible levels: biological, cognitively, emotionally and socially; furthermore, beloved people and the entire society are drastically affected.

2. Catherine Malabou does not think of the brain exclusively as an individual work in progress, but as a political model according to which we must rethink a multiplicity of interactions in which the participants exercise transformative effects on one another through the demands of recognition, of nondomination, and of liberty. See Catherine Malabou, *What Should We Do with Our Brain?*, trans. Sebastian Rand (New York: Fordham University Press, 2008), 31.

3. Ibid., 15–17. "Plasticity, though not altogether assimilable to rigidity, marks a certain determination of form [contrary to elasticity: elastic materials retain no imprint and, therefore, invite endless polymorphism, GM] and imposes a (very strict) restriction on the capacity for deformation, reformation, or explosion" (15). It is interesting to observe the range of meanings in the Greek language. The Greek *plassein* (πλάσσειν) has the following meanings. (1) It may mean to mold a three-dimensional material, resistant, and not always malleable; for sculpture, it opposes *graphein*, which, besides writing, means painting. (2) When used metaphorically, *plassein* means to imagine something in one's mind, or to form character by training or education. (3) It also means to fabricate: a story, fiction or myth, which is a meaning that does not emerge in the French language. In comparison to the French dialectic between fashioning and destruction (ibid., 78), the Greek meanings range from shaping to misshaping and forging.

4. This is not the right place to explain why, but let it suffice to say that Seneca, in his essay "*On the Shortness of Life*," provides sufficient argument why we should not be skipping the present moment when living our life.

5. Ibid., 63; see also ibid., 73.

6. One of the few exceptions is the philosopher and neuroscientist Alva Noë; see, for instance, Alva Noë, *Out of Our Heads: Why You Are Not Your Brain, and Other Lessons from the Biology of Consciousness* (New York: Hill and Wang, 2009).

7. Malabou mentions the dialogue between Changeux and Ricoeur in *What Makes Us Think*? as a good example of the two alternatives of reductionism and antireductionism (Malabou, *What Should We Do?*, 81f.). To add a representative of antireductionism in the analytic tradition of philosophy: Gilbert Ryle in his *The Concept of Mind.*

8. Catherine Malabou, *The New Wounded: From Neurosis to Brain Damage*, trans. Steven Miller (New York: Fordham University Press, 2012), 81.

9. The list of topics is not thereby exhausted. To be added is the relation between affectivity, in particular pleasure, and the experience of temporality, an experience that is distorted in the case of affective disorders. This topic goes beyond the current essay.

10. I do not choose to focus on Marcus Aurelius because I detected any Stoic traces in Malabou's work; on the contrary: she expresses strong opposition to any kind of determinism and to any possible extirpation of emotions, to which the Stoic therapy of emotions appears to amount.

11. On acceptance as a fundamental element in various new forms and expansions of cognitive behavioral therapies, see David Dozois and Aaron Beck, "Cognitive Therapy," in *Mindfulness and Acceptance*, ed. Steven C. Hayes, Victoria M. Follette, and Marsha M. Linehan (New York: Guilford Press, 2004), 26–56, and consider all contributions in Evan M. Forman and James D. Herbert, *Acceptance and Mindfulness in Cognitive Behavior Therapy. Understanding and Applying the New Therapies* (Hoboken, NJ: John Wiley, 2011). The founders of cognitive psychotherapies, Aaron Beck and Albert Ellis, have drawn upon Stoic philosophy, so the connection between Stoic philosophy and cognitive psychotherapy is definitely not new. What has lately been becoming a very influential and aspiring trend is the further elaboration on this multifarious connection and the fruitful and further promising collaboration between philosophers and historians of philosophy, on the one hand, and psychologists, on the other. See, for instance, Sellars's "Philosophy as Medicine: Stoicism and Cognitive Psychotherapy" (draft).

12. S. D. Hollon et al., "Prevention of Relapse Following Cognitive Therapy vs. Medications in Moderate to Severe Depression," *Archives of General Psychiatry* 62 (2005): 417–22. I am thankful to David Dozois for updating me about recent research in clinical depression, and also for helping me to understand the results of this study. What this study shows is that cognitive behavioral therapy alone is equally effective as medication, and that maintenance medication together with cognitive behavioral therapy is a very strong combination with regard to avoiding recurrence and relapse into depression. That said, we should not draw the conclusion that medication is not needed.

13. Let me explain why I speak of so-called feelings, emotions, and moods. Consider what cognitive psychotherapists label as "automatic thoughts"; for example, take the following among other "worthless core beliefs": "I am (worth of) nothing," a thought, which, if repeated by a person that has tendencies to, or past experience with, depressive episodes, necessitates depressive moods, a causation which shows that emotions are driven by beliefs and moods do not come out of the blue, but through

the intervention of particular thoughts. I think highly of different kinds of moods, including sad ones, but something that results from the above so-called thought is not a genuine mood, but a caricature of a mood. And let me add that the above so-called thought has absolutely nothing to do with what Marcus Aurelius prompts himself to think and reflect on in his *Meditations*, namely, that he himself and his entire life is worth of nothing when considered against the background of the whole universe and the whole of time. Marcus Aurelius's "I am nothing" is a reflection and not a caricatured thought. This kind of reflection does not necessitate depression. For, this is not an automatic thought. I think we have to seriously consider this grave difference so that we do not misunderstand Marcus Aurelius's meditations as depressive, a consideration that will also pave the way to providing further help to depressed people. That said, this is a task for philosophers and psychologists to fulfill: in no case should we hand over Marcus Aurelius's *Meditations* to people who are currently suffering from a depressive episode.

14. By Aaron Beck and Albert Ellis.

15. On the principles and the conception and nature of the body according to the Early Stoics, see Anthony Long and David Sedley, *Hellenistic Philosophy* (Cambridge: Cambridge University Press, 1987). It would be inappropriate to treat Marcus's *Meditations* as a theoretical treatise—for a very insightful reading about how we should read the work, instead, see John Sellars, "The Meditations and the Ancient Art of Living," in *A Companion to Marcus Aurelius*, ed. Marcel van Ackeren (Oxford: Oxford University Press, 2012), 453–64. Nonetheless, theoretical doctrines from the Early Stoa come to the surface here and there, and show that Marcus was acquainted with them.

16. Consider David Sedley, "Hellenistic Physics and Metaphysics," in *The Cambridge History of Hellenistic Philosophy*, eds. Keimpe Algra, Jonathan Barnes, Jaap Mansfeld, and Malcolm Schofield (Cambridge: Cambridge University Press, 1999), 383f. for antireductionism in the Stoics along these lines, and Anthony Long, "The Self in the Meditations," in *A Companion to Marcus Aurelius*, ed. Marcel van Ackeren (Oxford: Oxford University Press, 2012), 471. For a reductionist reading, consult Marcelo Boeri, "The Stoic Psychological Physicalism," *The New Centennial Review* 10 (2011): 105–32.

17. Compare 10.31 for the same analogy with fire. All translations of the Greek (Teubner edition) are mine. I have considered and especially benefitted from Gill's (2013) and Hard's translations.

18. See 11.5, in which Marcus reminds himself of the art to be good as his own art, which presupposes the knowledge of theorems of physics and ethics. The Stoic threefold division into physics, logics, and ethics, appropriated by Marcus for his own purposes, emerges in 7.67 and 8.13. In 10.31, he raises the expectation that the reason examines with the accuracy of physics all that life embraces. Therefore, Marcus explicates the theoretical presuppositions of his philosophical undertaking quite frequently, and, I read 1.17.17–20 in this light: there he makes clear that he did not wish to be concerned with physics and logic in the way that a theoretical treatise would do, and not that he thinks poorly of logic or physics.

19. See 7.68 and 11.5, and Sellars, "The Meditations and the Ancient Art of Living" on philosophy as the art of living in Marcus Aurelius and his predecessors.

20. To "live according to nature" is an old Stoic tenor. See Long and Sedley, *Hellenistic Philosophy*, 63, on how the earlier and later Stoics formulate the end in ethics and its relation to happiness. To live according to nature means to live according to the order in the universe (one meaning of *phusis*), which, for human beings, ends up meaning to live in accordance with the specific nature (another meaning of *phusis*) that the universe and divine providence allotted to us, namely, reason, in both its aspects, the rational and the political.

21. Consider the beautiful precision with which Plato sets up the model of craftsmanship in *Gorgias* 503d5–504a5. That said, whereas Plato's Socrates characterizes the craftsman's material as recalcitrant, matter is not recalcitrant according to the Stoics, but passive and malleable.

22. Marcus likes to repeat the dictum "All is judgment" that stems from Epictetus, the later Stoic philosopher who has exerted great influence on the Roman emperor: see 2.15; 12.8; 12.22; 12.26.

23. According to old Stoic theory, which Marcus shares, health, wealth, and fame and sickness, poverty, and social disgrace are neither good nor bad, respectively, but indifferent: when used for the sake of moral improvement, they are good; when they cause moral deterioration, they are bad. What is good in itself is virtue, and what is bad in itself is vice. That said, health, wealth, and fame are preferred indifferent items and sickness, poverty, and social disgrace are dispreferred indifferents. See Long and Sedley, *Hellenistic Philosophy*, 58 for passages on value and indifference.

24. See the end of 4.49: What appears to be a bad fortune is a good fortune when one endures it with noble spirit.

25. Exactly because the intellect is corporeal, the analogies between intellect and fire as well as between intellect and the stomach (see 10.31 on the latter analogy) are appropriate, though, they have their limits, since some fire can be extinguished by what falls upon it and some stomach may be destroyed by what it might receive, whereas this is not the case with intellect, which can accept and appropriate all possible events.

26. The combination of and the tension between Stoic and as some have detected Platonic elements in the *Meditations* is a bone of contention: see, for instance, Christopher Gill, "Marcus Aurelius' Meditations: How Stoic and how Platonic?" in *Platonic Stoicism, Stoic Platonism: The Dialogue Between Stoicism and Platonism in Antiquity*, ed. Mauro Bonazzi and Christoph Helmig (Leuven: Leuven University Press, 2007), 189–207. Marcus highlights mind's nature and power, and underlines its significant difference from other corporeal entities, but he does not end up wearing a Platonist hat, as I see it. For a Platonist conception of intellect as impassible, incorporeal, and, in this respect, separate, is not an option for and nowhere evidenced in the *Meditations*. Consider 5.26, according to which, when read through 6.32 and 12.30, intellect is part of the unified body. Though Marcus initially distinguishes body from soul (ψυχή) in 6.32, he then equates the latter with διάνοια, which shows that he did not preserve a strict distinction between soul and the intellect. Therefore, intellect is as corporeal as soul; for, the corporeality of the latter he has made explicit throughout the work. As for 12.30, he uses terminology from the Early Stoa, speaks of different bodies as ἰδίως ποιοῖς σώμασι, and interestingly refers to νοερὰ ψυχή and not νοῦς, which corroborates that he thinks of the intellect as material.

27. To translate Marcus's attitude into our own language in a different way: he maintains that all events are physical, but declines to reduce mental experiences and states to neurophysiological occurrences with which they are necessarily correlated. As I argued in section 2, it is not my intention here to prove that this is so in Marcus Aurelius, and, even less, to show that this is the case.

28. See 3.8 and 11.1 on the completeness of each present moment, in contrast to all actions and productions that need time to come to completion.

29. The perfect living being is a reference to Plato's *Timaeus*, and means the universe.

30. There are many things going on in this passage, which is characteristic of the way in which Marcus writes his *Meditations*, starting with cosmic teleology and divine providence, and including our deeply rational and political nature, and there are things we can be motivated to think about although we may disagree with Marcus and the Stoics in their fundamental tenets. In the future end-state he longs for, Marcus will not be skipping the present by either attempting to extend the present pleasure into the future or by desiring to attain the next pleasure in the future. He will not be wishing to prolong his present moment nor will he be postponing his fulfillment to the future. For, freed from the vicious circle of desire-satisfaction and desire-frustration, he will be resting in the present as complete in itself.

31. This is a distinction that goes back to the old Stoic doctrine, which Marcus Aurelius shares and expounds. See Long and Sedley, 65, on the passions according to the Stoics.

32. See 3.4; 3.16; 4.33; 5.8; 6.44; 8.7.

33. See 4.3, and compare 7.2.

34. I recall the Canadian pianist Janina Fialkowska, famous for her Chopin interpretations, who, in an interview, talked about the cancer of her left arm as *enabling* her to train this hand, a hand she had ignored until then, and, in this way, the older pieces she had played sounded a bit fresher then! One cannot love cancer or depression or whichever "bad luck" in themselves. One will love one's "fate" as one translates and transforms it.

35. See 8.49 for good advice: one should avoid value judgments and perceive whatever may be happening without evaluating it, an idea which cognitive behaviorist therapy has drawn upon later Stoic philosophy. The differentiation between thinking and facts and the understanding of thoughts as hypotheses and not facts are fundamental for the therapy of depression.

36. I only touch upon the second aspect of plasticity of time, and particularly the present time, in this context. I am exploring the plasticity of temporality in Marcus Aurelius's moral progress in a separate article.

37. "Plasticity Ought Not to Be Confused with the Mere Capacity to Act." Malabou, *What Should We Do?*, 48.

38. My critical point is restricted to the work I focus on; a generalization would not do justice to Catherine Malabou's thinking and concerns in other works.

39. I have in mind the analysis in Nietzsche's first *Untimely Meditation*, in which he draws upon Burckhardt's term "plastic power." I like to transfer his analysis of plastic culture to the individual's plastic transformation of their past, something Nietzsche does not focus on in this early work.

BIBLIOGRAPHY

Aurelius, Marcus Antoninus. *Ad Se Ipsum Libri XII*. Edited by Joachim Dalfen. Leipzig: Teubner, 1987.

———. *Meditations Books 1–6*. Translated by Christopher Gill. Oxford: Oxford University Press, 2013.

———. *The Meditations of Marcus Aurelius*. Translated by Robin Hard. Oxford: Oxford University Press, 2011.

Boeri, Marcelo. "The Stoic Psychological Physicalism." *The New Centennial Review* 10 (2011): 105–32.

Changeux, Jean-Pierre, and Paul Ricoeur. *What Makes Us Think? A Philosopher and a Scientist Argue about Ethics, Human Nature and Brain*. Translated by M. B. DeBevoise. Princeton, NJ: Princeton University Press, 2002.

Dozois, David, and Aaron Beck. "Cognitive Therapy." In *Mindfulness and Acceptance*, edited by Steven C. Hayes, Victoria M. Follette, and Marsha M. Linehan, 26–56. New York: Guilford Press, 2004.

Forman, Evan M., and James D. Herbert. *Acceptance and Mindfulness in Cognitive Behavior Therapy. Understanding and Applying the New Therapies*. Hoboken, NJ: John Wiley, 2011.

Gill, Christopher. "Marcus Aurelius' Meditations: How Stoic and how Platonic?" In *Platonic Stoicism, Stoic Platonism: The Dialogue between Stoicism and Platonism in Antiquity*, edited by Mauro Bonazzi and Christoph Helmig, 189–207. Leuven: Leuven University Press, 2007.

Hayes, Steven C., Victoria M. Follette, and Marsha M. Linehan. *Mindfulness and Acceptance*. New York: Guilford Press, 2004.

Hollon, S. D., R. J. DeRubeis, R. C. Shelton, J. D. Amsterdam, R. M. Salomon, J. P. O'Reardon, M. L. Lovett, P. R. Young, K. L. Haman, B. B. Freeman, and R. Gallop. "Prevention of Relapse Following Cognitive Therapy vs. Medications in Moderate to Severe Depression." *Archives of General Psychiatry* 62 (2005): 417–22.

Long, Anthony. "The Self in the Meditations." In *A Companion to Marcus*, edited by Marcel van Ackeren, 465–79. Oxford: Oxford University Press, 2012.

Long, Anthony, and David Sedley. *Hellenistic Philosophy*, 2 vols. Cambridge: Cambridge University Press, 1987.

Malabou, Catherine. *The New Wounded: From Neurosis to Brain Damage*. Translated by Steven Miller. New York: Fordham University Press, 2012.

———. *What Should We Do with Our Brain?* Translated by Sebastian Rand. New York: Fordham University Press, 2008.

Malabou, Catherine, and Adrian Johnston. *Self and Emotional Life: Philosophy, Psychoanalysis and Neuroscience*. New York: Columbia University Press, 2013.

Nietzsche, Friedrich. "Unzeitgemäße Betrachtungen" In *Friedrich Nietzsche, Kritische Studienausgabe, Vol. I*, edited by Girgio Colli and Mazzino Montinari. München: De Gruyter, 1988.

Noë, Alva. *Out of Our Heads: Why You Are Not Your Brain, and Other Lessons from the Biology of Consciousness*. New York: Hill and Wang, 2009.

Rist, John. "Are You a Stoic? The Case of Marcus Aurelius." In *Jewish and Christian Self-Definition 3*, edited by B. F. Meyers and E. P. Sanders, 23–45. London: SCM, 1982.

Robertson, Donald. *The Philosophy of Cognitive Behavioral Therapy: Stoic Philosophy as Rational and Cognitive Psychotherapy*. London: Karnac, 2010.

Ryle, Gilbert. *The Concept of Mind*. Chicago: University of Chicago Press, 1949.

Sedley, David. "Hellenistic Physics and Metaphysics." In *The Cambridge History of Hellenistic Philosophy*, edited by Keimpe Algra, Jonathan Barnes, Jaap Mansfeld, and Malcolm Schofield, 353–411. Cambridge: Cambridge University Press, 1999.

Sellars, John. "The Meditations and the Ancient Art of Living." In *A Companion to Marcus Aurelius*, edited by Marcel van Ackeren, 453–64. Oxford: Oxford University Press, 2012.

———. "Philosophy as Medicine: Stoicism and Cognitive Psychotherapy." Draft, 2012.

Seneca, Lucius Annaeus. "De Brevitate Vitae." In *Seneca, Moral Essays, Volume II*, 286–355. Loeb Classical Library 254. Cambridge, MA: Harvard University Press, 1932.

Chapter 13

"The Still Missing People"

Thinking the Affective Work of Art in the Work of Gilles Deleuze, through Catherine Malabou's Concept of Plasticity

Meadhbh Mcnutt

In discovering that the brain is constituted, not of a central authoritative core but of billions of neurons (interdependent yet not without their own competences), the authority of the autonomous individual is confronted with an inconsistent identity, grounded in a paradoxical pairing of decentralized consciousness with the unified appearance of the body. What is at stake in this discrepancy in light of current neuroscientific discourses, and where does this situate the subject in the production and affective experience of art? My inquiry looks to the continuously altered *neuronal self* in delineating a conjointly altered conception of art. My focus will be Catherine Malabou's work on the topic of neuroplasticity and the precedence of wonder in the history of metaphysics. Catherine Malabou has previously compared the neuronal self's formation to that of a sculpture. In a text entitled *What Should We Do with Our Brain?* she writes: "It is precisely because—contrary to what we normally think—the brain is not already made that we must ask what we should do with it, what we should do with this plasticity that makes us, precisely in the sense of a work: sculpture, modelling, architecture. What should we do with this plastic organic art?"[1]

The image of the plastic brain as a work of art makes for a convincing analogy but how does this analogy inform our understanding of the artwork in turn? My question pertains to the idea of an imminent form of resistance specific to the artwork, one which poses a critique to the concept of the subject. At the heart of this criticality, as interpreted in aesthetics and critical theory thus far, is the capacity to wonder. Wonder here acts as a placeholder for praxis in a society wherein action cannot exist outside the clutches of repressive ideologies and instrumental reason. It remains an implicit condition in various theoretical accounts of artistic production and experience. For Malabou, wonder is the primary affect which accounts for the very structure of the plastic brain, providing mediation between intrusion and integration. As her later work on the neurological accident suggests, this primary affect is far from indestructible. With the simultaneous necessity and destructibility of wonder in mind, my aim is to complicate the pervasive understanding of artistic experience as one of straightforward harmonious autoaffection, or philosophical reflection. In comparing brain plasticity with the concept of plasticity in its "native" artistic field, I do not intend to reduce art—a complicated social and historical practice—to an essentialist concept propped up by the empirical sciences. As Malabou clarifies in her study of plasticity: temporality and historicity are inseparable from the plastic, biological formation of the brain. It is from the perspective of such an intertwinement that I approach the idea of art practice as a production of affect, outlined in Deleuze and Guattari's text, *What Is Philosophy?* My study of the role of the affective subject in art ultimately leads me to Malabou's work on the disaffected subject—an investigation which unsettles the pervading notion of the artwork as a mirror for the subject.

FROM PROTO TO EXPERIENTIAL PLASTICITY

What Should We Do with Our Brain?, a text that lays much of the groundwork for Malabou's following studies on disaffection, navigates through the various forms of plasticity, namely: developmental plasticity (the creation of neuronal connections), modulational plasticity (the reorganization of neuronal pathways in response to life experience) and reparative plasticity (the reorganization of neuronal pathways in salvaging brain functions following a brain lesion/neurodegenerative disorder). Brain plasticity can be defined as the brain's capacity to simultaneously impose and receive form through the reorganization of neuronal pathways in response to sensory stimulation.[2] Neuronal pathways are reorganized in the process of *apoptosis*, that is, the controlled death of an organism's cells as part of its natural growth and development. Cell death acts as the "sculptor's chisel,"[3] eliminating that which the brain's neuronal network deems unnecessary. In this way, the parts of the brain are united only through a specific form of dissonance which calls into question the harmony of the neuronal network. The brain is an agent insofar as it imposes its own form, yet such an imposition is dependent on the capacity to receive form from exterior stimuli. Recalling Spinoza's proposition that affect emerges between affection and action, Malabou's concept of plasticity occurs at the point of contention between homeostasis and self-generation. The former, operating at the level of protoself, accounts for the tendency of a system "to maintain internal stability" through the coordinated response of its parts to stimuli that disturbs its normal functioning. Interpretive functions which allow the brain to be affected are also responsible for the brain's ability to preserve its own form. The brain is plastic by virtue of a process of self-affection, whereby phenomena are firstly interpreted by the brain, altering its state. This alteration is then interpreted in turn as a kind of "autoaffection." In experiencing exterior stimuli, consciousness produces an obscured neuronal image of itself. It is therefore impossible to distinguish the two domains of mind and brain absolutely. Consequently, according to Malabou, there is no complete synaptic theory of personality. We are fluent in the language of the self to the extent that our fluency disallows us from seeing the self in an uninterrupted state. She does however conclude that this kind of connectivity and representational activity is such that neuronal networks are "disposed toward" meaning.[4] What is in question in respect to the translation of neuronal to mental is the lawfulness of such a disposition. Malabou here proposes a fourth form of plasticity, in the transition from neuronal to mental that occurs as an ontological leap. This fourth form acts as a kind of bind between protoplasticity and experiential plasticity,[5] accumulating in an act of "self-generation." Self-generation, the tendency of the brain to rebel against its own process of coordination, alters the physiological form of the brain through *apoptosis* in what Malabou calls, an explosive response. What follows is a morphological transformation from the physiological to the biographical. Counter to this, flexibility can be interpreted as a process of submission in the sense that the brain is available to receive form but cannot autonomously alter itself in spite of its homeostatic tendencies. To be flexible rather than plastic is to refuse any destruction to the brain's system of coordination—a refusal of plasticity's self-generating capacity. Flexibility has become an ideological trope in current capitalist societies that conveniently dovetails uncritical notions of neuronal creativity. This trope enforces neuronal stasis and represses experiential plasticity by conflating flexibility with creativity, "protoself" with biographical self.

An artwork is critical, or plastic, when it facilitates an experience in which the subject recognizes the tendency of form to resist itself. Such a criticality breaks through the façade of harmony—posing a threat to the authority of the subject. How does affectivity intervene in this scene? Throughout his aesthetic works Deleuze has attempted to unpack this question, turning ultimately to cerebral matters in his later collaboration with Félix Guattari, *What is Philosophy?* Thinking art through Malabou's concept of plasticity in conjunction with Deleuze's work may seem a strained endeavor, given that the idea of plasticity is derived from a contradictory relation between mind and nature in Hegel's philosophy that presents thought as "negated nature"—a conviction at odds with Deleuze's affirmative plane of immanence. Despite these theoretical disparities surrounding negativity, Malabou makes clear in *Self and Emotional Life* that the Deleuzian corpus allows for a reading of the conflict between metaphysics and neurobiology, raising important questions of the precedence of wonder in the history of metaphysics. It is worth mentioning that Deleuze and Guattari's final chapter in *What Is Philosophy?*, "From Chaos to the Brain" tends to migrate toward negativity, in matters of cerebral cell death and in distinguishing the three disciplines of art, science, and philosophy.[6] Deleuze even goes so far as to acknowledge the neuronal process of apoptosis in the following passage: "And individuation in the cerebral state of affairs . . . does not have the cells themselves for variables, since the latter constantly die without being renewed, making the brain a set of little deaths that puts constant death within us."[7]

The influence of both Baruch Spinoza's foresights of protoneurology and René Descartes's metaphysics of presence plays out in Deleuze's framing of art in terms of affective production along a plane of composition. *What Should We Do with Our Brain?* outlines a becoming-world of the brain whose form is comparable to an artwork in its contradictory tendency to persist in resisting its own given form. In *What Is Philosophy?*, an artwork is a speculative form of resistance that persists in annulling its given figures, casting a shadow of the "people to come," that is "world-people" or "brain-people."

FORM AND RESISTANCE

Throughout Malabou's study, two factors appear to be of importance to plasticity's political implications, that is, affect and resistance—key concerns shared by Deleuze and Guattari, in their discourse on art. For Deleuze and Guattari, designating affect as a modification of a wider ontology is imperative in discussing matters of the brain or mind. The depth of Spinoza's influence on Deleuze's work elucidates this move—an influence which extends to Malabou and the neurologists cited in her work, particularly Antonio Damasio. Spinoza's work has taken on new significance as a way to radically rethink the subject following discoveries concerning the biological stratification of emotion, and its indispensable role in rational thought. A distinction is needed here between the primary status of affect and its corresponding emotions in order to avoid using these intertwined yet specific terms interchangeably. Affect is defined by Spinoza as a modification of the "conatus," which produces a feeling or emotion in the finite mode through which it is experienced. On distinguishing the emotional from the cognitive product of affect, Malabou explains: "This is because the encounter does not trigger any faculty or sense or logical structure; it touches—and thus reveals—the very feeling of existence."[8]

To demonstrate how Malabou's concept of "plasticity" and Deleuze's use of the term "affect" extend and contextualize one another, I will turn to

Malabou's tracing of Spinoza's theory of autoaffection in Deleuze's work in her text, *Self and Emotional Life*. During a lecture on Spinoza's *Ethics* (January 24, 1978), Deleuze specifies that Spinoza's "affect" refers to the Latin *affectus*, and that his use of "affection" corresponds with *affectio*.[9] *Affectus* implies a modification elicited by an encounter. According to Spinoza, we act when something internal or external to us occurs that "follows our nature" or of which we are an adequate cause. We are acted on insofar as something "follows from our nature of which we are only the partial cause."[10] Action, then, increases the force of existing in that the idea I have of myself is affirmed by that which "follows my nature," the effect of which is felt as an active and joyous affection. Spinoza describes affects as primary alterations constitutive of the body, of which the mind has a necessarily confused idea.[11] He offers a mapping of consciousness in the idea of the composite body. In general terms, a composite body, of which the human body is an example, constitutes an individual composed of a number of individuals. The human body is not only composite itself; it is also comprised of composite parts which have been constrained by external bodies, and united by a fixed pattern of communication between each part. Pivotal to this manner of communication is the concept of the conatus that specifies an ontological disposition in which "each thing, in so far as it is in itself, endeavors to persist in its own being."[12] What does it mean for something to persist "in its own being?" Being, firstly, implies a certain state of existence; that this being is "in itself" suggests that the persistence of such a state is essential to the thing. How then could we identify the conatus, the perseverance of a specific existing state, as an essential tendency of a finite individual, one which continuously decomposes and recomposes in encountering external bodies? How can this unfixed state account for a self-affirming essence? Although Spinoza lays claim to the immutability of substance,[13] he challenges the substantialist claim that form is transitory while substance remains fixed. All bodies are animate insofar as they "either move or are at rest."[14] An essential state is thereby not equivalent to a fixed state. Rather, the essence of a body is a relation of its parts to one another, a negotiation analogous to that of a human society. Spinoza describes this agreement in the scholium as follows:

> When a number of bodies, whether of the same or of different size, are so constrained by other bodies . . . that they communicate their motions to each other in a certain fixed manner, we shall say that those bodies are united with one another and that they all together compose one body or individual, which is distinguished from the others by this union of bodies.[15]

Those individuals which make up the body agree in communicating rest and movement but also reconfigure in response to encounters with external bodies, transforming with the severity of these heteronomous disruptions. Such a principle of motion is partly influenced by Descartes's *Principles of Philosophy*, as we can see from similarities between Spinoza's adoption of "strive" and Descartes's earlier use of "tend," notable in Spinoza's critique of *Principles*: "By striving for motion (*conatum ad motum*) we don't understand any thought, but only that a part of matter is so placed and stirred to motion, that it really would go somewhere if it were not prevented by (*impediantur*) any cause."[16]

Given that the body is composed of, and continually destabilized by reciprocal exchanges among internal and external bodies, its complex composition mutilates the very idea it has of itself. A rupture occurs in the idea of the body, between the tendency to strive for self-preservation and the destructive processes sedimented in its very form. Indeed, the defining line between destruction and construction blurs significantly. Encounters with unanticipated exterior bodies are structurally incorporated into an individual's form, composing a kind of historicity of disfigurement. As Deleuze summarizes in *Expressionism in Philosophy*: "Reason distinguishes action from opinion (*doxa*) as an expression of essence and an affirmation of the capacity to be affected."[17]

The divine power of the conatus is understood in Deleuze's *Practical Philosophy* according to a schema of intensity, through which the necessary simultaneity of the capacity to affect and to be affected is implied. The conatus according to Deleuze is a kind of autoaffection by which essence modifies itself. Affect encounters itself as a "spacing," a line of flight from and within itself—a mapping of essence. Such a surface is what Deleuze calls the plane of immanence: a mapping, "connectable in all its dimensions" that persists through its own modification.[18]

Malabou likens Deleuze's conception of autoaffection to artistic creation, the ideal solicitation of which would be "painting or imaging itself." But what occurs in this transition from autoaffection, an internal solicitation, to pure affect externalized in the work of art? In his reading of Descartes's *Passions of The Soul*, Deleuze identifies the external signs of the passions in the flexes around the face as primary, visible points of affective surface-creation. These appear as "expressions of eye and face, changes in color, trembling, listlessness, fainting, laughter, tears."[19] Descartes's animal spirits, which cause movement-action in the limbs and muscles, are distinguished from those of the passions of the soul. The latter causes micromovements to surface at the face from a concealed source, such that it becomes a surface upon which affects are inscribed as points of impact. We can think of affects as prepersonal, in that they produce a subject through their own mapping activity. In the disruption of the regulated movements of the face, being itself comes to occupy the visage of the subject. As pure being surfaces, so too does the fundamental affect—a "degree zero"[20] as Malabou describes it—which amounts to wonder. Wonder (*admiratio*), in other words, signifies the emotional capacity to be amazed, concerned, or to inquire.

Spinoza's concept of the conatus is relevant to Malabou's contentions on form insofar as both resist the idea that form is transformed, while substance remains unchanged. A subsection in *The New Wounded* entitled "'They have all been dead,' or how Spinoza was right" revisits Damasio's *Spinoza Avait Raison*, gesturing toward another truth unearthed from the *Ethics* that Damasio failed to notice: that any living body can "be changed into another nature entirely different from its own."[21] The very nature of affect implies the possibility of a completely novel existence, wherein the body does not recognize itself as a historical being. Destructive plasticity thus involves an irrevocable transformation and, with it, an unknown kind of suffering. Although Deleuze does not make any explicit statements about the destructibility of wonder or the capacity be affected, he recognizes the need for "another level of possible variation," which recognizes the unfixed nature of the capacity to be affected, or to wonder.[22] An instability of this kind transforms the very structure of the conatus, constituting a second type of "expressive variation." It is unclear just how this observation informs his later aesthetic works. He does however note a sense of self-detachment in the flattened architecture of the subjective "I." As a plane of immanence, the I becomes "icon."[23]

THE ONTOLOGY OF THE ARTWORK

There is a critical aspect to this thinking through of detachment in the artwork, when artistic experience gestures beyond a self-touching in the sense of an immediate self-reflection or closure of subjectivity upon itself. Affect takes shape in the artwork as an independent sonorous block surpassing the material locus. But how can this work be at once sensory and independent of the senses of its recipients? Deleuze and Guattari liken sensation to contemplation or contraction. The plane of immanence constitutes spatial encounters, both interior and exterior to the body. In *What is Philosophy?*, Deleuze and Guattari define chaos in regard to the plane of immanence as the "infinite speed of birth and disappearance."[24] Chaos is not so much disorder as the untraceability or indistinguishability of order, as in the case of an unexpected rupture or accident. Protolife develops solely out of chaos through a process of contraction. Contraction in this context involves at once a drawing together of the separated and a folding of something onto itself—an encounter of immanent difference. The power of form to touch and modify itself, resisting its own form in doing so, gives way to surface-creation, or protolife. This is precisely the work of consciousness, a self-reflective contraction of previous alterations that sediment in the body. History and potential are projected within its form. We can thus think of sensation as nothing more than the contraction of the "vibrations of the stimulant on a nervous surface or in a cerebral volume—what comes before has not yet disappeared when what follows appears."[25] In first person, the self maps itself as the second person, encountering itself only as a map or spacing in the third. This movement of becoming permeates the very corporeality of the brain. Art as a nonhuman becoming of affect, shares an affinity with the infinite speed of chaos and thus a resistance against the stagnant nature of opinion (*doxa*). I refer here to opinion as derivative "functions of lived experiences,"[26] that is to say, propositions correlating a specific affect with a generic subject. Deleuze and Guattari's concept of opinion shares similarities with Malabou's thoughts on flexibility. Both imply a regulative malleability—a failure to impose form and disrupt regulative states. Designating art as a sensory framing of chaos, Deleuze and Guattari write, "Art takes a bit of chaos in a frame in order to form a composed chaos that becomes sensory, or from which it extracts a chaoid sensation as variety."[27] Art is necessarily bound up in sensation for Deleuze and Guattari. This necessary relation does not imply a flat denial of the conceptual content of the artwork but rather emphasizes the significance of the work's experiential character which must always engage with that from which it separates (the empirically given) in order that it preserve its capacity for resistance. The varieties ("chaoid sensations") that art takes from chaos are not representations of sensory experience, nor are they a reproduction of the sensory organ but on the contrary, a being of sensation that follows its own individual logic. Rather than enveloping chaos as opinion aims to achieve, varieties maintain aspects of chaos in the interest of restoring the infinite on a plane of composition. The use of the term "chaos" is pertinent in the context of contemporary capitalism in which the idea of crisis looms large, and social progression is often conflated with political stasis and risk aversion. This sort of conflation would be an example of *doxa* for Deleuze, the notion of which he compares to an umbrella whose function is to provide complete shelter from chaos. Art plays a destructive role in tearing through the umbrella, only to frame the eternal speed of chaos that appears through the slit—to render it sensory without submersion. We can envisage this analogy in Nabokov's side door: "While a few pertinent points have to be marked, the general impression I desire to convey is of a side door crashing open in life's full flight, and a rush of roaring black time drowning with its whipping wind the cry of lone disaster."[28] The risk that this roaring black time may resurface and erase the sedimented contractions of history is ever present in forms of political trauma. Malabou describes the neurological accident as "hopeless, unpredictable . . . an accident that cannot be integrated by the psyche, that cannot make sense for it, that cannot form a moment of a personal history."[29]

Flexibility offers little in the way of a solution. With precarious forms of labor now the norm, old boundaries between public and private, work and play, are gradually dissolving. Flexibility, the tendency by which the brain preserves itself, is encouraged as a form of agency or self-actualization. It thus becomes a method of self-policing, a necessary condition by which the neoliberal logic of crisis currently operates. The threat of irrevocable affective detachment—the "meaningless of wounds"—is perhaps its silent enforcer.[30] Both Malabou and Deleuze call for resistance, that is: the ability to say no. Deleuze often evaluates a work by its ability to "stand up by itself."[31] Resistance in this sense implies an integrity of the work's particularity. It is the task of art to make sensible the elementary power of resistance in the genesis of form, by probing and deconstructing the very concept of art in infinitely many ways. In this sense, art is always double in character—primordial and projective.

Throughout *The New Wounded: From Neurosis to Brain Damage*, Malabou stresses the impossibility of separating the effects of political trauma from that of organic trauma. Is this absence of emotion, as in the case of a brain lesion or a response to trauma, a kind of affect in itself due to its modification of the "feeling of existing," or is it precisely a loss of affect? Is the subject who does not recognize itself as a subject who suffers, still a subject? Malabou outlines the possibility of hetero-heteroaffection in the following passage: "A heteroaffected subject is still an affected subject. A hetero-heteroaffected subject is "disaffected." Most of the time, the impairment of emotional processes produces an indifference that coincides purely and simply with a disability to wonder."[32] The brain owes its historical character to the capacity of affect, to encounter itself through wonder. The historical and social aspect of art is similarly inseparable from its affective capacity. Is the disappearance of affect, then, a disappearance of history itself, in the work of art as in the brain? Malabou concludes *The New Wounded* with the following claim: "The history of being itself consists perhaps of nothing but a series of accidents which, in every era and without hope of return, dangerously disfigure the meaning of essence."[33]

If we know little else of the essence of art than that it is temporal or historical, we may understand it as a practice of speculation and retrospectively signified disfiguration. Art persists through negation, paving the way for retrospective signification. The relationship between form and history is both synchronic and diachronic in this context. As Malabou puts it,

> On the one hand, the coincidence between formation and disappearance of form is diachronic: a past form cedes place to a new form, and one thus changes identity or "self" in the course of time. On the other hand, the coincidence between form and disappearance is synchronic: the threat of the explosion of form structurally inhabits every form.[34]

THE WORLD CLOCK: A SYSTEM ORBITING THE ACCIDENT

But how does this negation occur in the creative process? At what point does a creative act of resistance foster an affinity with chaos? Deleuze put this down to impulses of control and flight. This conflict between chaos and control—a recurring trope in the history of art—is evident in the work of Lars von Trier, whose films produce contours of chaos through strict systems of control. The *Dogme 95* manifesto, coauthored with Thomas Vinterberg in 1995, outlines rigid limitations to filming methods with the alleged aim of cinematic purification and the implicit aim of producing unique transgressions within a well-constructed system—unprecedented lines of flight.

Psykomobile #1: The World Clock was an installation conceived by Lars von Trier in the year following *Dogme 95*. Dispersed throughout nineteen individually named rooms in Copenhagen's Kunstforeningen, a group of fifty-three actors embodied incongruous characters over a period of fifty days (three hours per day, six days per week). In a hefty manual, Von Trier and scriptwriter Niels Vørsel laid out character sketches for each actor, providing limited details of their lives, surroundings and mutual relationships. Each actor would stay true to the short description of their character's

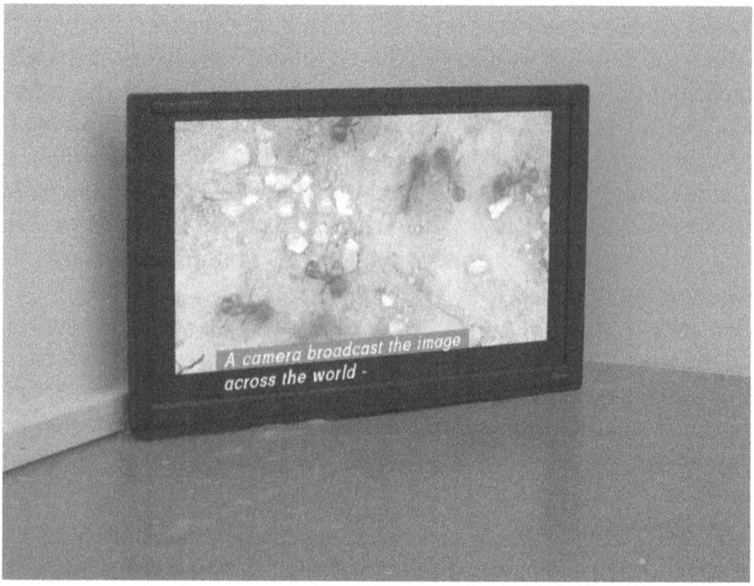

Figure 13.1. Melissa Dubbin and Aaron S. Davidson, *Myrmomancy (Ants Gathered and Transmitted for Lars Von Trier's The World Clock)*, 1996/2014. Single channel video, color, and sound (Source: Courtesy of the artists and Untilthen, Paris).

psychological disposition but had the freedom to explore whatever was left unsaid.[35] Footage of the "living installation" shows a flurry of improvization. As actors roam between rooms at their own will, it becomes unclear who is participant and who is observer in the congested exhibition space. A light changes every so often in a room, signaling the actors to alter the mindset of their characters. The dynamics of rules and transgressions at the center of this work are revealed at the entrance hall where a large screen displays a live video feed of ants crawling on arid New Mexican soil. The video feed is a work in itself by Melissa Dubbin and Aaron S. Davidson, *Myrmomancy* 1996/2014, a live transmission via satellite links (see figures 13.1 and 13.2). The gathering of ants for Von Trier's *World Clock* was described by the artists as a form of myrmomancy, an "occult practice of reading the future through observation of the behaviour of ants." In this case, the occult practice is refined to a near-exact science—a controlled experiment. When four ants crawl into one of the squares, a change of light is triggered in its counterpart room. The cognitive and emotional states adopted by the actors are entirely contingent on the movement of a group of ants on the other side of the world.

Von Trier's theater is born out of strict direction and refinement only to be thrown to the whims of a different species. Actors are governed by an oblivious colony whose social and psychological structure remains impenetrable to them. *The World Clock* puts forward a controlled scheme that participants

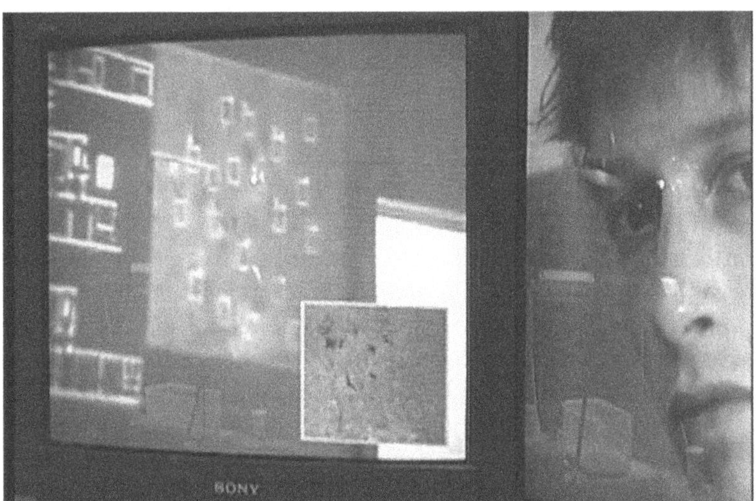

Figure 13.2. Melissa Dubbin and Aaron S. Davidson, *Myrmomancy (Ants Gathered and Transmitted for Lars Von Trier's The World Clock)*, 1996/2014. Single channel video, color, and sound (Source: Courtesy of the artists and Untilthen, Paris).

adhere to yet ultimately engulf in their own unforeseen improvizations, only to return back to the regulative guidelines upon a change in light. This reciprocal relation between threat and regulation is schematized by the composition displayed on the live screen, where unpredictable swarms and fixed zones of governance are unlikely coconspirators. Homeostasis and self-generation echo in the discretionary set of rules that are created, rules asserted only to be exposed to frequent contestation and transformation. Von Trier appears to construct a parody of the spectator who overdetermines the conceptual content concealed within the artwork; however, a focus on the intention of the artist as one of one-upmanship disregards the wider tendency throughout Von Trier's work to grapple with the powerlessness of the symbolic in the face of physiological contingency. Despite the strictness of its manual, the *World Clock* is subjected to the power of chance and accident. Von Trier's films often challenge the viewer to witness the aftermath of an accident in which the character has experienced a complete loss of affect. This confrontation with the disaffected allows the audience to envisage a world in which there is no reattachment of affect to itself and in doing so, to consider the unaffected subject's place in the world as a kind of nonsubject. The neural trauma, a forced entry which the psyche struggles to integrate into an existing symbolic economy, speaks to the processes at work in *The World Clock*. Its participants encounter the opaque behavior of a distant ant colony as a concealed impetus. Abstract signals (traffic lights) are armed with a disruptive force that shapes the entire installation. The work comes across as an absurd parody—an inexhaustive cycle of mediation, disruption, and self-alteration mapped through the banal corpus of the Kunstforeningen. This sort of absurd sentiment reappears throughout von Trier's later work. As the fox in *Antichrist* would later tell us, having consumed his own flesh: "chaos reigns."

Here we return to Malabou's proposition that the history of being is perhaps little more than a series of accidents, one after the next, recognized and integrated into the psyche retrospectively. *The World Clock* renders its own form at once historical and ludicrously accidental. A tension upholds this system, which adheres to, and intervenes on its own self-preservation—at once echoing and throwing into question the harmonious unity of the subject. In this sense, the audience is invited to rethink the consolatory function of the artwork as a pure reflection of the humanity of the subject. As the randomness of the fluctuating lights begin to take effect, the actors find it increasingly difficult to distinguish their own emotions from those dictated by the overall system. Viewers intermingle with actors. One actor claims that certain switches of light "changed lives." The power by which the *Psykomobile* keeps turning is dependent on its ability to incorporate disfiguration into the very structure of its own impetus. The installation becomes its own animal—a living organism, parodying its own claim to freedom in imposing a structure of chaos. Yet the potential for a new form of praxis is felt in this parodic freedom. The specific resolutions offered up by both fixed opinion and unbounded crisis are thrown in the air, suspended. The work thus casts a speculative "shadow" through which an altogether different logic threatens to emerge.

ART THAT CASTS A SHADOW OF THE "STILL MISSING PEOPLE"

Art practice, a social form of labor, is far from reducible to a straightforward expression of the subject. However, its complex relationship with the subject and autonomy, further complicated by the philosophical implications of neurological research, cannot be disregarded. Much of this research is instrumentalized by the culture industries, glossed with an optimistic veneer of creativity and served up as proof of inherent cerebral virtues. The more commonplace blind optimism toward the empirical sciences, the more is at stake in a critical engagement with ideas of the neuronal dimension of artistic experience. Malabou has made a case for the importance of attempts by neurologists to question the ideological foundations of their research (Changeux, Jeannerod, and Damasio are just a few examples). Transforming the concept of plasticity while also respecting its merit as a scientific concept in clinical practices and empirical research, she insists, "Creating resistance to neuronal ideology is what our brain wants, and what we want for it."[36] Malabou's proposition of a "reasonable materialism" moves beyond a reductionist-antireductionist dichotomy, allowing for an insight into the destructibility of wonder and, in turn, a revaluation of the precedence of wonder in artistic and philosophical discourses. Confronting this precedence of the primary affect, Malabou calls into question the certainty of the subject, and its place in historical narrative. If as Deleuze asserts, the disciplines of art, science, and philosophy endure as points of contention between the dogma of opinion and complete disappearance into chaos, between flexible submission and cool detachment, wonder appears to conjoin all three disciplines in its inclination toward chaos and accident. But what of a complete submergence into chaos, wherein even the most rudimentary identity cannot recognize itself? It is because wonder *is* destructible that art must face the potential of its own eradication. Art can only persist in submerging itself in nonart—in anticipating a disfigurement in the mechanism which admits the production of affect. This nonart in which art must find itself again is not the passive sensation of sorrow which still contains potential for change. Art retains its aptitude for resistance in producing affects that present not only the possibility of passivity but the possibility of their own destruction. Just as the disaffected subject exists at the border of subjectivity and nonsubjectivity, so must art face the border of nonart. Deleuze presents this nonart as indistinct from chaos in *What Is Philosophy* and as the nihility of the Sahara desert in *The Logic of Sensation*. In either case, nonart must be present at every moment of an artwork's becoming. While describing a Francis Bacon painting, Deleuze writes,

> A Sahara, a rhinoceros skin: such is the suddenly outstretched diagram. It is as if, in the midst of the figurative and probabilistic givens, a catastrophe overcame the canvas. It is like the emergence of another world. For these marks, these traits, are irrational, involuntary, accidental, free.[37]

In flattening and clearing its given figures, art casts the shadow of a new people, a "still missing people." The still missing people are today those people that lie at the borders of subjectivity, the "ontological refugee."[38] The impossibility of concern, of a mediation through the feeling of existence, disrupts the structural integration of the threat of explosion and results in what Malabou adeptly describes as a "coolness," or detachment. Art distinguishes itself from this irrevocable detachment in prolonging the promise of chaos. Thus complete destruction, as that which "does not come to pass," must remain implicit to the very concept of the artwork. This is what gives art its speculative character. The work of art, and the brain as a work of art, transforms itself in resisting itself. It emerges in the interchange between system and lawlessness, malleability and explosion, composition and chaos.

NOTES

1. Catherine Malabou, *What Should We Do with Our Brain?*, trans. Sebastian Rand (New York: Fordham University Press, 2008), 7.
2. Definition of brain plasticity derived from J. C. Ameisen, *La Sculpture Du Vivant: Le Suicide Cellulaire Ou La Mort Créatrice*, translated by Malabou in *What Should We Do with Our Brain?*
3. Ibid., 19.
4. Ibid., 62.
5. Ibid., 69.
6. These disciplines are no longer distinct if the brain is subsumed in chaos. "Now, if the three Nos are still distinct in relation to the cerebral plane, they are no longer distinct in relation to the chaos into which the brain plunges." Gilles Deleuze and Félix Guattari, *What Is Philosophy?* (New York: Columbia University Press, 1994), 218.
7. Ibid., 216.
8. Catherine Malabou and Adrian Johnston, *Self and Emotional Life: Philosophy, Psychoanalysis, and Neuroscience* (New York: Columbia University Press, 2013), 59.
9. Gilles Deleuze, *Lectures on Spinoza* at Vincennes, 1978.
10. Benedictus De Spinoza, *Ethics*, trans. Edwin Curley (London: Penguin Classics, 1997) bk. 2. prop 2.
11. Ibid. bk. 2. prop. 28, 135.

12. Ibid. bk. 3. prop. 6, 159.

13. "Substance is, by its nature, infinite, immutable, indivisible, and so forth." Ibid. bk. 2. prop. 10, 121.

14. Ibid., 126.

15. Ibid. bk. 2. l. 3. a2, 126.

16. Benedictus De Spinoza, *Principles of Cartesian Philosophy* (Philosophical Library/Open Road, 2014) 3.56.

17. Gilles Deleuze, *Expressionism in Philosophy: Spinoza* (New York: Zone, 1990), 226.

18. Gilles Deleuze and Félix Guattari, *A Thousand Plateaus: Capitalism and Schizophrenia* (Minneapolis: University of Minnesota Press, 1987), 45.

19. Malabou, *Self and Emotional Life*, 172.

20. Ibid., 45.

21. Catherine Malabou, *The New Wounded: From Neurosis to Brain Damage* (New York: Fordham University Press, 2012) 202.

22. "For the capacity to be affected does not remain fixed at all times and from all viewpoints," Deleuze, *Expressionism in Philosophy*, 222.

23. Ibid., 175.

24. Deleuze and Guattari, *What is Philosophy*, 118.

25. Ibid., 211.

26. Ibid., 174.

27. Ibid., 206.

28. Vladimir Nabokov, *Lolita* (New York: Penguin Classics, 2000), 168.

29. Malabou, *Self and Emotional Life*, 216.

30. Malabou, *New Wounded*, 53.

31. Deleuze and Guattari, *What Is Philosophy*?, 178.

32. Ibid., 77.

33. Malabou, *New Wounded*, 91.

34. Malabou, *What Should We Do?*, 71.

35. Jan Simons, *Playing the Waves: Lars Von Trier's Game Cinema* (Amsterdam: Amsterdam University Press, 2007), 46.

36. Malabou, *What Should We Do?*, 77.

37. Gilles Deleuze, *Francis Bacon: The Logic of Sensation* (Minneapolis: University of Minnesota Press, 2005), 100.

38. Term taken from Malabou, *New Wounded*, 24.

BIBLIOGRAPHY

Print

Damasio, Antonio R. *Looking for Spinoza: Joy, Sorrow, and the Feeling Brain.* Orlando, FL: Harcourt, 2003.

Deleuze, Gilles. *Expressionism in Philosophy: Spinoza*. New York: Zone, 1990.

———. *Francis Bacon: The Logic of Sensation*. Translated by Daniel W. Smith. Minneapolis: University of Minnesota Press, 2005.

———. *Spinoza: Practical Philosophy*. San Francisco, CA: City Lights, 1988.

Deleuze, Gilles, and Felix Guattari. *A Thousand Plateaus: Capitalism and Schizophrenia*. Minneapolis: University of Minnesota Press, 1987.
———. *What Is Philosophy?* Translated by Hugh Tomlinson and Graham Burchell. New York: Columbia University Press, 1994.
Grosz, Elizabeth. *Chaos, Territory, Art: Deleuze and the Framing of the Earth*. New York: Columbia University Press, 2008.
Malabou, Catherine. *The New Wounded: From Neurosis to Brain Damage*. New York: Fordham University Press, 2012.
———. *Ontology of the Accident: An Essay on Destructive Plasticity*. Translated by Carolyn Shread. Cambridge: Polity Books, 2012.
———. *The Future of Hegel: Plasticity, Temporality and Dialectic*. Translated by Lisbeth During. London: Routledge, 2004.
———. *What Should We Do with Our Brain?* Translated by Sebastian Rand. New York: Fordham University Press, 2008.
Malabou, Catherine, and Adrian Johnston. *Self and Emotional Life: Philosophy, Psychoanalysis, and Neuroscience*. New York: Columbia University Press, 2013.
Nabokov, Vladimir Vladimirovich. *Lolita*. New York: Penguin Classics, 2000.
Rancière, Jacques. "The Monument and Its Confidences; or Deleuze and Art's Capacity of 'Resistance.'" In *Dissensus: On Politics and Aesthetics*, edited and translated by Steve Corcoran. London: Continuum, 2010.
Descartes, René. *Passions of the Soul*. Translated by Stephen H. Voss. Cambridge, MA: Hackett, 1989.
Simons, Jan. *Playing the Waves: Lars Von Trier's Game Cinema*. Amsterdam: Amsterdam University Press, 2007.
Spinoza, Benedictus De. *Ethics*. Translated by Edwin Curley. London: Penguin Classics, 1996.
———. *Principles of Cartesian Philosophy*. New York: Philosophical Library/Open Road, 2014.

Film/Television

De Udstillede (The Exhibited). Directed by Jesper Jargil. 2000. Danish Cultural Developmental Pool.
Kulturoperatørene. Directed by Erik Jacobsen, Mona Levin, and Marta Norheim. 1996. Copenhagen: NRK (Norsk Rikskringkasting AS).
Antichrist. Directed by Lars Von Trier. 2009. Nordisk Film Distribution.

Performance

Von Trier, Lars. *Psykomobile #1: The World Clock*. Kunstforeningen Copenhagen. 1996. Performance.

Web

Deleuze, Gilles. Lecture Transcripts on Spinoza's Concept of Affect. Goldsmiths, University of London. Available from http://www.webdeleuze.com.

Chapter 14

Diagnosing the Sociopolitical Wound

Frantz Fanon and Catherine Malabou

Sujaya Dhanvantari

> The determination of psychic disturbances—their definition, their clinical picture, and their therapy—is always contemporaneous with a certain state or a certain age of war.[1]
>
> —Catherine Malabou

> There have never been in the history of humanity, in absolute numbers, so many inequalities, so many cases of malnutrition, ecological disaster, or rampant epidemic (think, for example, of AIDS in Africa and of the millions of people we allow to die, and thus kill!).[2]
>
> —Jacques Derrida

This chapter proposes an encounter between Frantz Fanon's mid-twentieth-century concept of the colonized psyche, and the current philosophical investigations by Catherine Malabou into the sociopolitical dimensions of neurological disease. Rereading Fanon today will give his revolutionary clinical work on mental illness another sense that blends the biological brain with its oppressive context. Even as Fanon will not have anticipated Malabou's concepts of cerebrality and destructive plasticity, we could now find in that work an interpretation of *the neuropathological trauma of the colonized subject*.[3]

In sum, this chapter will consider Fanon and Malabou's link between psychic trauma and political oppression. Specifically, it will read Fanon's critique of the negative effect of colonialism on the biological body and psyche in relation to Malabou's effacement of the border between biological and political violence in her concept of the new wounded. Further, it will link the traumatic event of colonization to contemporary sociopolitical oppression. Finally, this study will find that a spectral colonial trauma haunts the psychic pathologies of today's oppressed subjectivities.

FROM FANON TO MALABOU

The figure of the racialized and colonized subject appears in Fanon's early writings: his first published article "The North African Syndrome" and book *Black Skin, White Masks* in 1952. There, Fanon concludes that European colonialism is the cause of psychic disturbances in his immigrant, minority patients in Lyon, France. In 1951, Fanon was first living and working as a clinical psychiatrist, France was still a colonial power.[4] As a witness the inner workings of the French psychiatric hospital, Fanon observed the racial typecasting of the North African patient by his physician colleagues.

The North Africans whom Fanon treated had "phantom illnesses"—or "pseudo-illnesses"—related to their negative experiences of racialization and colonization in France. The North African patient was afflicted with an obscure malady. In effect, the patient's suffering begged for an analysis that would make sense of this "invisible" wound. But their condition remained illegible, and consequently the patient received no proper clinical therapy. Fanon observed that even in the absence of specific indicators of illness, these patients were troubled by a feeling of imminent "death" ("Doctor, I'm going to die!").[5] Accordingly, Fanon concluded that a diagnosis would have to include their lived experience in systemically oppressive environments. In all of Fanon's writings, the colonized subject is constituted as a biological body and psyche negatively changed by exposure to oppressive situations. Fanon concludes the colonial situation perversely creates a "morbid" colonized personality that presents self-destructive feelings and behavior.

In 1952, the year Fanon published "The North African Syndrome," a critical analysis of his professional interactions with racially diverse patients prior to and during his residency at the *Hôpital de Saint Alban* in the Lozère Region, he had come under the influence of Dr. Francois Tosquelles, a Catalan refugee who employed the revolutionary new therapeutic techniques that integrated communality into the strictly carceral psychiatric institution.[6] Upon leaving Saint Alban, Fanon brought that critique to his own medical practices in Blida-Joinville and Tunis.[7]

The French psychiatric institution played a prominent role in colonial expansion over the course of the nineteenth and twentieth centuries. For Fanon, the medical profession in particular, and the history of French colonization in general, caused the colonized subject to experience an "iatrogenic" form of illness ("The truth is that colonization, in its very essence, already appeared to be a great purveyor of psychiatric hospitals").[8] *Fanon's analysis can now be linked to Malabou's concept of negative plasticity*. Malabou asks if a traumatic event can form a new identity whereby "neuronal plasticity destroys synaptic connections."[9] Reading Fanon with Malabou enables an analysis of the catastrophic event of colonization as a trigger for the subject's inner drives of destruction, generating a *destructive metamorphosis*. The colonized patient's specific diagnostic profile must then identify the psychic pathologies related to the destructive effect of colonial oppression.

That is why Fanon's colonized subjects present a *class* within Malabou's category of the new wounded. This paradigm centers on those suffering from cerebral lesions (from the accidental wound to Alzheimer's to newly diagnosed chronic illnesses), but also includes the victims of war and other types of political oppression: all those cases were either rejected, ignored or unknown in traditional psychoanalytical practice.[10] It will be argued that the psychic disturbances identified by Fanon can today be classed alongside those traumatic illnesses that are now unified under one clinical profile.[11] Fanon's critical analysis of the psychic wound of colonization constitutes a fertile ground for extending the parameters of the new wounded into the colonial past. A continuum of violent histories appears on the scene of this encounter between Fanon and Malabou, which exposes an unexplored relation between the psychic illnesses contemporaneous with the age of European colonization and the new profiles of traumatized identities.

Malabou develops a general theory of trauma in her research on the neurological sciences. She states that the victims of sociopolitical violence experience similar symptoms to those with neuronal lesions.[12] Fanon did not of course know of the language of brain plasticity. For his medical degree at the University of Lyon, however, he studied the psychiatric symptoms in patients with a hereditary neurodegenerative disease linked to "Friedrich's ataxia" (57). In his dissertation, Fanon argues that even in this "congenital" neurological disorder, a related mental illness would form itself only *in time and space*. He concludes that neither a strictly neurological nor a strictly psychiatric approach could alone determine the nature of mental illness that unfolds in the sociocultural context. Even the tendency to draw a simple neuropsychiatric correlation had to be disputed if it excluded the sociocultural factor.[13] But the idea that the cerebral cortex was conceptually linked to the development of neurological condition in the oppressed, remained central in his later work at Blida-Joinville, where he observed patients decerebralize[d] by their colonial situation.[14]

In her deployment of the neurological sciences to elucidate psychic trauma, however, Malabou does not seek to reproduce that historically divisive moment when neurology correlated itself with psychiatry, while excluding the sociopolitical environment. Rather, she constructs a neuropathology of the oppressed. In effect, Fanon's early critique of the exclusion of the sociocultural factor from the diagnostic assessment invites Malabou's concept of cerebrality that integrates it. Today's oppressed subject appears at that border between the neurological sciences and the sociopolitical world.

FANON'S CLINICAL PROFILE OF THE COLONIZED SUBJECT

A Pathology of Oppression

Fanon elaborates a general paradigm of psychic disorders that appear within situations of racialization and colonialism, first evident in his psychiatric work and lived experience in postwar colonial France. In "The North African Syndrome," Fanon argues that doctors operating within a colonial institutional model systemically racialized the North African. Even if no physical lesions could be verified, Fanon argued that these patients presented with *pseudo-illnesses* associated with the negative experience of expatriation.

In the typical doctor's visit described by Fanon, the white bourgeois physician disregards the symptoms of the ailing "Arab" patient, labeling him "a liar, a malingerer, a sluggard, a thief." In the "Arab" with a "belly ache," the French doctor sees "a-man-who-doesn't-like-to-work."[15] The malingerer symbolizes a common narcissistic neurotic in the diagnostic stereotype that integrates colonial racism.[16]

Likewise, Fanon witnesses the clinical practice of infantilizing the racialized patient within the discipline of psychiatry, which he interprets as a doctor's ethical mistake.[17] Talking to the "Arab" as an equal, Fanon says, would require an empathetic relation. Further, physician researchers such as Dr. Leon Mugneiry openly racialized the North African as essentially pathological.[18] To counter prolific racism in the psychiatric profession of the colonial period, Fanon employs Dr. Heinrich Meng's theory of the *situational* diagnosis, arriving at the conclusion that the North African in France experiences "a living death."[19] This form of death in the living subject is an effect of the oppressive situation. The racism present inside and outside the institution results in the psychic suffering of the North African without leaving any *lesional* trace. In his first article, Fanon puzzles over his early observation that the nonlesional expression of psychic disturbances is found in his racialized patients suffering from the situational effects of expatriation.[20]

Importantly, Fanon defines the impact of institutional racism on the psyche of the patient who does not display any physiological symptom yet suffers. In effect, Fanon discovers that the patient's experience of racial exclusion triggers forms of psychological distress that results in a type of psychosomatic illness. Hence, Fanon began his critique of colonialism by classifying the

"North African Syndrome" as a bona fide condition that resulted from the damaging effects of living in situations of expatriation.[21] Fanon more clearly elaborates on the space of psychic suffering of the racialized subject in *Black Skin, White Masks*.

Fanon and the Nervous Personality

In "The Lived Experience of the Black Man," chapter five of *Black Skin, White Masks*, the racialized psyche feels its oppressive weight in the first person, conveyed through the various metaphors of physiological ailments, including those of blood deficiency, chills, nausea, bleeding, and excessive nervous energy. Denied equality, Fanon writes, "I get upset; a frightening rage makes me anemic."[22] Elsewhere, the subject trembles in the presence of the gaze encircling him. Although he attributes his own involuntary movements to climate-related cold chills, the white child sees them as angry tremors. The archaic processes of racialization isolate him further from others in the present, ending in fits of nausea: "I approached the Other . . . and the Other, evasive, hostile, but not opaque, transparent and absent, vanished. Nausea."[23] Later, feeling sequestered by this inescapable enclosure within the circle of racialization, he says, "Shame. Shame and self-contempt. Nausea."[24]

The "body schema" disintegrates, "attacked in several places, collapsed, giving way to an epidermal racial schema."[25] His biological body is shattered and reconstructed inhumanly into his "epidermis," which becomes the site of a racial binary. When the other cannot recognize him as an equal, he experiences himself as both distended and torn apart. "In the train, it was a question of being aware of my body, no longer in the third person but in triple. In the train, instead of one seat, they left me two or three." The body's psychic overextension beyond its physical limitations is accompanied by a paradoxical internal hemorrhaging, "Peeling, stripping my skin, causing a hemorrhage that left congealed black blood all over my body."[26] Being unrecognized by others leaves him feeling torn apart—internally reenacting the terror of the Middle Passage.

The racialized subject experiences an intensified, but vacuous, emotional pain permeating a physiological state of suffering with no evidential trace, But, he refuses to inoculate himself, "I refused, however, any affective tetanization," understanding that the inoculation would quiet the pain, while having a destructive power.[27] He sees himself exploding ("Nothing doing. I explode. Here are the fragments put together by another me").[28] Then, in spite of his shattered self, he accesses a hidden reserve of energy: "I can feel that familiar rush of blood surge up from the numerous dispersions of my being. I am about to lose my temper."[29]

But the new identity built out of the fragments of the self/other relation and remolded into "another me" is the effect of a destructive change. The past that it can no longer identify is experienced as a death haunting psychic life. No inoculation can cure the patient of what it no longer possesses, even if it is possessed by it as by a phantom limb. The oppressed subject can only now reconstruct its prior identity by reflecting on it, but without any associative experience. Fanon's subject grasps at the experience of slavery from outside living memory, attempting to name the force of political trauma that is reactivated in the child's words. He poetically reconstructs that deep colonial wound in an attempt to "diagnose" the pain that repeats itself through the child's slights. That past however cannot be relived or possessed. But it can be accessed through the processes of "thought": "I transported myself on that particular day far, very far, from myself, and gave myself up as an object."[30] Through the process of objectively analyzing the colonial event, the subject is able to grasp the depth of its present state of trauma.

Fanon senses that the colonial psyche is ready to explode. He metaphorically condenses and internalizes that outward expression of the explosive event into the figure of the nervous system. The racialized subject feels its oppression in fits of nervous energy, trembling, and mental distress ("as for me, I was a nervous wreck, shaking at the slightest alert").[31] Elsewhere, the subject suffocates. Indeed, death returns in this text but within the circle of life through the figures of suffocation, asphyxiation, even incineration: "All this whiteness burns me to a cinder." The subject experiences itself oppressively as a disappearance inside the living body, or as a *living figure of death*. "The fire had died a long time ago, and once again the Negro is trembling."[32]

The fact that he is still breathing, despite the effect of asphyxiation having turned him against himself, illustrates that his subjectivity has been oppressed, forcing him to transform himself into the racialized figure of the colonial world. He feels the absence of the deserted identity as the missing "limbs" of an "amputation."[33] Nonetheless, the state of a "living death" is not experienced as a rebirth, but as a continual relapse of that death in life, one that is felt to be an agony over a permanent absence. The oppressiveness is metaphorically constructed into the shape of a "circle" enclosing the subject within a fixed and unbroken concentric space and time of closure: "I was walled in."[34] It is an imprisonment lived out inside the psyche and biological body of the oppressed ("I am a prisoner of the vicious circle").[35]

Reading psychic disorders contextually, Fanon emphasizes the fact that colonial experience damages psychiatric health, permanently imprinting subjectivity with irreversible histories of violence and exploitation. Fanon visualizes the slave ship traversing the Atlantic Ocean from within the standpoint of mid-twentieth-century Lyon. This body in present life is disjointed just as the time is "out of joint"—the time of the archaic past is revived within the living present. This passage on Fanon's experience of alienation discloses the historical trauma unleashed by present-day racism.[36] In effect, the childhood of black interracial relations constituted a site of violent subjective formation, now recapped on the streets of Lyon, where a white child points and says, "Look, a Negro."[37] But the child's words are cut from the past, and now wield the force of that trauma.

Fanon views the analysis of the silenced spectral horror of slavery as a critical task for the colonized subject. A new *memory* of slavery would create an attachment to the ancestral past; that past could then act as a resource for a positive transformation. Fanon searches for the historical origins of the child's old words that make him bleed, and then decides to keep the old wound open, to give himself the time to assess the depth and nature of it: a self-diagnosis.

THE "NEW WOUNDED"

Neurological Disorders: A New Diagnosis of Oppression

Fanon includes the social, economic, and political conditions of the colonial context in his psychiatric studies. This chapter now proposes that Fanon invites Malabou's new concept of the cerebral state of shock blurring lesional and nonlesional disorders within the category of the new wounded.

As stated, Fanon's diagnosis takes the colonial wound into account. The psychic wounds of colonial oppression appear in the metaphor of physical "crippling" ("The crippled soldier from the Pacific war tells my brother: 'Get used to your color the way I got used to my stump. We are both casualties.'").[38] Fanon returns the political to its material counterpart in the biological body and psyche, inviting Malabou's blending of neuropathic and sociopathic disorders. She declares that "the traumatic event is the psychoneurotic event," displacing Freud's "neurosis" into the interdisciplinary study of psychoanalysis and neurology.[39] The concept of cerebrality demands an analysis of the impact of the traumatic event on the formation of the psychic wound, whose effect is shared by the victims of both brain lesions and political oppression.

Fanon's diagnosis of the colonial wounded can be brought into relation with the discoveries in the neurosciences and the emerging list of chronic illnesses. Malabou writes, "The 'new wounded' constitute an emergent phenomenon, then, to the extent that this category also refers to subjects who suffer from disturbances that had yet to be identified during Freud's time. For example, one might adduce several recently discovered disorders:

obsessive-compulsive disturbances, hyperactivity syndrome with attention deficit disorder, or any of the illnesses identified by the 'disabilities movement.' "[40]

The enlarged category includes the previously mute psychic traumas experienced by the victims of all forms of catastrophe: first, it adds the patients of accidental head wounds, together with those of degenerative brain diseases such as Alzheimer's and the incurable cases of psychic impairment ("schizophrenics, autistics, epileptics, victims of Tourette's syndrome");[41] and secondly, it introduces the victims of chronic illnesses bearing no physiological lesions. The "unification" of these "disparate clinical profiles" suggests conceptualizing the biological expression of observationally similar personality changes. Hence, Malabou blurs the boundary between the victims of brain lesions and those who suffer emotional disturbances related to psychological stresses.[42] Accordingly, for Malabou, the category of the new wounded covers "every patient in a *state of shock* who, without having suffered brain lesions, has seen their neuronal organization and psychic equilibrium permanently changed by trauma." She adds: "Such patients also suffer, in particular, from an emotional deficit."[43]

Furthermore, Malabou concludes that "the behavior of subjects who are victims of trauma linked to mistreatment, war, terrorist attacks, captivity, or sexual abuse display striking resemblances with subjects who have suffered brain damage. It is possible to name these traumas 'sociopolitical traumas.' Under this generic term, one should group all damage caused by extreme relational violence."[44] Based on the clinical observation of convergent behavior patterns, Malabou argues that the identical expressions of emotional disturbance across profiles suggest the porosity of the border between "organic trauma and political trauma." Even if the former is the cause and the latter the consequence of psychic disturbance, "the same *impact of event* is at work, the same *economy of the accident*, the same relation between psyche and catastrophe."[45] It is thus possible, even likely, that Fanon's colonized subject suffered from a type of *posttraumatic stress*. We could now return with greater compassion to the expatriate patients who felt themselves dying in colonial hospitals. We might find that their suffering is linked to the emergence of contemporary forms of illness.

Indeed, the brain in the concept of cerebrality constitutes the site of subjective formation, where psychic shocks and disturbances—even if they do not leave lesions—are felt by the biological brain and body as a psychological stress caused by those external pressures that have an unpredictable or accidental character. Malabou introduces the biological brain from the neurological sciences to the study of political trauma: "The work of contemporary neurologists has helped me to discover *the impossibility of separating the effects of political trauma from the effects of organic trauma*."[46] In effect, the shock or disturbance of sociopolitical violence is registered by the biological brain's emotional receptors: "Even in the absence of any patent wound, we know today that any shock, any especially strong psychological stress, or any acute anxiety, always impacts the *affective brain*—this unrecognized part of the psyche."[47] Accordingly, the new subject—the victim of this *destructive metamorphosis*—displays an emotional deficit.

This negative behavior can be interpreted as the immanent expression of the intrusive external event. It is a shared effect spanning the variations in lesional and political traumas, globalized through the results of neurological study. Fanon's colonial wound linked to Malabou's sociopolitical wound presents the traumatic experiences of today's oppressed subject within cerebral etiologies that bear a deep history.[48]

Destructive Plasticity in Fanon's Colonized Subject

"The specificity of the traumatic event thus inheres in the *metamorphic* power. The traumatic event, in a certain sense, invents the subject," writes Malabou. In other words, the negative plastic power triggered by the event creatively sculpts the new subject as a form of its own self-annihilation. *Fanon will not have anticipated the metamorphic power of destruction in his figure of living death.* For Malabou, the self-destructive drive is immanent in the living figures of death, illustrating that the death drive is effectively at work in the traumatized identity. She writes, "The deserted identities of cerebrality, *living figures of death*, emerge precisely as representatives of the tendency to annihilation and destruction that psychoanalysis has always failed to bring to light."[49]

Malabou argues that the psyche's innate potential for "deflagration or explosion of every form" has been overlooked in both psychoanalytic and neurological studies.[50] Plastic forms of creativity, namely, the powers to mold and be molded, have been the focus of the new interest in brain plasticity. But that third form—destructive plasticity—offers a way to comprehend the violent form of the "dying" unconscious, or the brain inwardly directed toward its own death, witnessed in the terrible expressions of disaffection. For Malabou, the metamorphic power of destruction discloses the organism's immanent powers of self-annihilation in situations where emotional stress overwhelms psychic life.

Even if the wound possesses a "sculpting power," "we are quite far from the sculptural paradigm of the 'beautiful form,' " says Malabou. The sculpting of the wound can only be thought of as generative to the extent that it triggers a negative-formative plastic power of creative destruction ("creation through the destruction of form").[51] The capacity to form a new identity while dislocating the "old" is an effect of the wound's plasticity, but it cannot be thought without the third self-annihilating drive. The oppressed subject survives by plastically becoming the form of its destruction, possessed by a tendency for annihilation.

The violence of closure that triggers the self-annihilating drive appears in Fanon's metaphor of an unbroken narrative of whiteness: "The white man is all around me; up above the sky is tearing at its navel; the earth crunches under my feet and sings white, white. All this whiteness burns me to a cinder."[52] Later, "When I opened my eyes yesterday I saw the sky in total revulsion. I tried to get up but the eviscerated silence surged toward me with paralyzed wings. . . . I began to weep."[53] passivity and indifference turn the oppressed subject inward, separating him from his affective relation to others. "Paralysis," Malabou writes, indicates "an inability to be touched affectively, which is the sign that one has been 'touched'—that is wounded."[54] Through the traumatic experience, the subject feels inexorably drawn to the mortal state of destructibility, whose outer expression is the catastrophic event. But the fact that Fanon's subject weeps indicates that a positive self-transformation is still possible from within the somatic and psychic state of inertia. Hence, the experience of evisceration paradoxically generates a positive form.

Malabou's concept of destructive plasticity is evidently at work in much of Fanon's writings. We can now see how the colonial wound sets in motion a horrific sculpting power that compels the colonized psyche to desert itself and plastically remold a new identity, which comes "closer to death" through the various states of emotional disaffection. If it survives, it will plastically sculpt a dark, violent detour to its own death, directing itself further inward into "its own disaffection" to the point that its affective faculties atrophy and it becomes increasingly solitary.[55] The subject of political oppression activates an innate self-destructive drive, which is "immanent to the organism itself"; for the colonized subject that drive is triggered by the extreme relational violence of colonization and the subsequent oppressive closure of the colonial world.[56]

It is therefore possible for Fanon's speaker to feel entirely destroyed, invisible, eviscerated, nauseated, and burned to ashes or likewise paralyzed or crippled. Emotional disaffection and the innate self-destructive forces are thus definitively knotted together. Malabou asserts that a certain state of psychic disaffection replicates the effects of the cerebral lesion. According to the neurological research Malabou cites in *The New Wounded*, it has become clear that the brain undergoes normal destructive changes when exposed to abnormal situations.[57] In Fanon's writings we see that colonization and oppressive closure are such abnormal situations that trigger the destructive drive immanent in the biological brain and psyche. The oppressed subject overcome with a sense of disaffection feels himself dying in Fanon's clinic even as he cannot name the "senselessness" of the external acts of political violence that his brain receives as a series of cerebral shocks.

For Malabou, the brain lesion is the paradigmatic cause of psychic trauma. Through the brain wound, the contours of political trauma emerge. Both lesional and political wounds send the traumatized subjectivity into a state of emotional coolness or indifference, drawing it toward its immanent state of "organic inertia" away from the life drives. Malabou explains that the psychic condition of emotional disaffection can be traced to a deficit of emotional receptors in the brains of those suffering either lesional or political trauma. Fanon's traumatized subject is thus inexorably pulled into a constricted space in which it feels itself to be "vanished." That state of disaffection ingrains a sense of "disaffiliation" in the social context—the subject feels itself organically disintegrating, while also feeling split from the social environment. Certainly, Fanon's racialized patients could also be classed as such victims.[58]

The oppressed subject cannot feel the passionate engagement of the life drives; in fact, both the outward expression of aggression and the inner self-destructive force could be interpreted as "diverted expressions of indifference or organic inertia."[59] In the desertion of identity, the brain loses its emotional receptors. Accordingly, the subject appears as indifferent. That is why the North African patients are in the waiting room at Fanon's clinic. A few years later in Algeria, Fanon will illustrate that indifference in the Manichean situation, which opens onto a similarly dark psychic space of inner disaffection, triggering the annihilative powers.[60]

It is noted however that the speaker in chapter 5 of *Black Skin, White Masks* still "weeps" even if psychically paralyzed. That is a sign that a formative plastic power asserts itself from within the destructive mode. Likewise, the North African patient visits the clinic: an opening into the brain's formative plasticity—the reception and donation of form—paradoxically at work in the wounded subjectivity of colonial history.[61]

CONCLUSION: THE POSTTRAUMATIC STATE—A COLONIAL INHERITANCE?

Is colonial trauma a spectral force at work in new psychic wounds? Certainly the recent posttraumatic diagnosis that multiplies morbidities suggests a radical effacement of "sense" in all forms of sociopolitical violence, and the DSM-III[62] has recently concluded a globalized clinical profile of the "universal" traumatic experience that blurs the distinctions between military and civilian battlefields, including social wars, while also conclusively effacing entrenched cultural, racial, and social distinctions.[63] Does a uniform pathology of political trauma indicate the invention of a globalized form of colonization, a new traumatic force running through the "brain" of the subject?

Colonial history would become a time and space that reappears at the border between new situations of political violence and the biologically responsive brain, activating explosive, self-destructive personality changes in the present. New diagnoses are required to fight what Malabou calls a redoubled "war against hermeneutics" in a political climate that prevents the powers of scientific, psychoanalytic, and philosophical interpretation from offering therapeutic possibilities. Fanon's diagnosis of the French North African must invite the redefinition of psychic trauma as a globalized *posttraumatic condition*.[64]

To begin, the neuropathological redefinition of the "traumatic etiology" that mixes situations of war with peace through the discovery of the posttraumatic stress disorder, or PTSD (in 1980), deepens the notion that specific "battlefields" in both military and civilian life pass through the brain of the subject, having the blunt force effect of a "bullet" or explosive device even in the absence of a cerebral lesion. By 1994, the ubiquity of PTSD was recognized in the link "between the traumatic factor and the experience of fear, impotence, or horror provoked by an objective event that may not have to do with war."[65] Any type of trauma could now be thought to provoke conditions of severe psychological stress leading to radical transformations in psychic life.

Much of the new research in neuropathology analyzed by Malabou finds that the pervasive examples of emotional disaffection and coolness indicate a new state of psychological trauma with global dimensions. Malabou adds that today's emotional coolness of subjectivity mirrors the "disaffection" of social, political, and economic systems. Contemporary social and political organizations could thus also be diagnosed as disaffected, reproducing a time and space of psychic destruction at the level of the social and political institution.[66] That analytical profile suggests that new forms of oppression are in the process of being invented.

Derrida points out in the epigraph of my chapter that the globalized instances of hunger, ecological disaster, and epidemic are the new forms of political violence typical of "the so-called age of globalization" that witnesses the "letting die" of ever greater numbers minus an awareness of political terror. Malabou concludes that the proliferation of these new sites of violence has the effect of a "shock" that resists interpretation.

Malabou diagnoses the sociopolitical wound that migrates across the geopolitical borders of the postcolonial state via the objective data that constructs "a new worldwide typology of psychic illness."[67] This suggests also that colonialism metamorphosized into new, mobile forms of oppression nebulous instances or authors dangerously dissimulate the traumatic event as a state of nature: "The dissimulation of the reason for the event is the new form of the event. The increasingly radical effacement of the distinction between accident and crime, between disastrous incidents and war, the multiform presence of the absence of any responsible instance or author makes *the natural catastrophe of contemporary politics* into a daily occurrence."[68] The parallel profile of the victim of a natural disaster and the victim of war violence indicates the absence of *sense* in the latter. In the absence of analysis, perpetrators are disregarded, and politics assumes a face of coolness or a state of disaffection. In Malabou's words, "We have entered a new age of political violence in which politics is defined by the renunciation of any hope of endowing violence with a political sense."[69]

The senselessness of the dissimulated political catastrophe is mirrored in the oppressed subjects who express the senselessness of their own suffering through psychoneurotic behavior. Without reason or the ability to choose or will in the absence of affectivity, the subject loses the capacity to temporalize itself. Malabou continues: "The effacement of sense, of course, does not only occur in countries at war; it is *everywhere*. It constitutes the new face of the social—bearing witness to an emergent, *globalized* psychic pathology that is identical in all cases and all contexts."[70] The politics that masks its intention with the face of the natural catastrophe dissimulates responsibility and neutralizes its purpose both in situations of war and in general across the sociopolitical landscape of globalization. Malabou's call for a new hermeneutics thus addresses the need for psychoanalytic, scientific, and philosophical interpretations to make sense of political violence. From Fanon to Malabou, from the colonial to the sociopolitical wound, colonial trauma has metamorphosed and migrated into a post-traumatic state.

NOTES

1. Catherine Malabou, *The New Wounded: From Neurosis to Brain Damage*, trans. Steven Miller (New York: Fordham University Press, 2012), xiv.

2. Jacques Derrida, "Autoimmunity: Real and Symbolic Suicides: A Dialogue with Jacques Derrida," interview by Giovanna Borradori, *Philosophy in a Time of Terror: Dialogues with Jürgen Habermas and Jacques Derrida* (Chicago: University of Chicago Press, 2003), 121–22.

3. Anibal Quijano and Walter D. Mignolo illustrate in decolonial thought that cultural destruction in the colonies produced colonial wounds that inscribed the *racial differential* onto non-European, non-white colonized peoples (Mignolo, 340). The violence of the colonial paradigm that racialized and excluded others resulted in a trauma that began with the conquest of the Americas, one which continues into the present. See Walter D. Mignolo, "Delinking: The Rhetoric of Modernity, the Logic of Coloniality and the Grammar of De-coloniality," *Globalization and the Decolonial*

Option, ed. Walter D. Mignolo and Arturo Escobar (New York: Routledge, 2010), 340, and Anibal Quijano, "Coloniality and Modernity/Rationality," 24, in the same collection.

4. France was a colonial power until 1962. In "Interview with Catherine Malabou" with Brenna Bhandar and Jonathan Goldberg-Hiller, Malabou says, "Among the colonizing ex-powers, France is one of those that will have taken the longest to reflect on its past. Postcolonial studies have met with success there only recently, and the French often deny the colonial trauma." She implies that the parallel temporalities of colonial and postcolonial France disclose a lack of acknowledgement of historical oppression in the present. See "Interview with Catherine Malabou," in *Plastic Materialities: Politics, Legality, and Metamorphosis in the Work of Catherine Malabou*, ed. Brenna Bhandar and Johnathan Goldberg-Hiller (Durham, NC: Duke University Press, 2015), 295–96.

5. Frantz Fanon, "The North African Syndrome," *Toward the African Revolution*, trans. Haakon Chevalier (New York: Grove, 1967 [1964]), 13.

6. In "Frantz Fanon 1925–1961," *History of Psychiatry* 7, no. 28 (1996): 489–97, David Macey writes that Fanon's most formative clinical experience was at Saint Alban in 1952, where he prepared for and received the *Medicat des hopitaux psychiatriques*, qualifying him to work as a professional psychiatrist in the French system. Importantly, Fanon's mentor at Saint-Alban was François Tosquelles, a Catalan refugee from Franco's Spain, who abandoned the "carceral" psychiatric model for one that "was designed to be what the British tradition calls a therapeutic community" (493).

7. Ibid., 493–95.

8. Frantz Fanon, *The Wretched of the Earth*, trans. Richard Philcox (New York: Grove Press, 2004 [1961]), 181.

9. Catherine Malabou, *The New Wounded*, 166.

10. Ibid., 9–10.

11. In assessing the entirety of Fanon's psychiatric practice, Jean Khalfa ("Fanon and Psychiatry," *Nottingham French Studies*, no. 1 [2015]: 52–71) states that although Fanon progressed from a position that opposed psychiatry to neurology to a stage of experimentation with neuropsychiatric treatment, his main focus was the "sociotherapeutical approach" that emphasized the essential relationship between culture and mental illness (53). Likewise, Khalfa argues that Fanon's dissertation specifies that the neurodegenerative disease is an innate hereditary condition, he development of mental illness in the affected subjects is essentially related to external sociocultural factors.

12. Peter Gratton, "Malabou's Plasticity of the Real," *Speculative Realism: Problems and Prospects* (London: Bloomsbury Academic, 2014), 188.

13. Khalfa, "Fanon and Psychiatry," 53–55.

14. Fanon writes, "The events of Algeria are the logical outcome of an abortive attempt to decerebralize a people." See *Toward the African Revolution: Political Essays*, trans. Haakon Chevalier (New York: Grove Press, 1967), 53.

15. Frantz Fanon. "The North African Syndrome," in *Toward the African Revolution: Political Essays*, trans. Haakon Chevalier (New York: Grove, 1967 [1964]), 4–7.

16. David Macey in "Frantz Fanon 1925–1961" states that both "The North African Syndrome" and *Black Skin, White Masks* "are products of Fanon's hostile

encounter with the work of the so-called Algiers school of psychiatry and his early clinical experience working with the immigrants (mainly Algerian) attracted to Lyon by the chemical and textile industries" (492). That school professed the civilizational theories of white supremacy, and produced professional manuals on clinical psychiatry including articles that pathologized the North African until the 1970s (492).

17. In "The North African Syndrome," Fanon criticized the French doctors who addressed their racialized patients condescendingly in the second person: *Où as-tu mal?* In Frantz Fanon, *Le colonisé en question: Le "syndrome nord africain" Pour la révolution africaine: Écrits politiques*. (Paris: Francois Maspero, 1969), 10.

18. Ibid., 12. Fanon cites Mugniery's position on the question of French citizenship.

19. Ibid., 13–17. Also, Ranjana Khanna in *Dark Continents: Psychoanalysis and Colonialism* (Durham, NC: Duke University Press, 2003) writes, "Fanon cites Dr. E. Stern's summary of Meng's theory. Stern is dismissive of the relevance of Meng's criteria for the Algerian Muslim" (175). Even as Fanon offers counterpositions for Stern's interpretations of Meng's criteria, Khanna points out that "it is not clear with whom Fanon identifies"; Khanna cites Memmi's observation that "the personal pronouns in this essay suggest an identification with the French" as opposed to his later writings that distinguish a colonized "we" from the French colonizer. In spite of the ambiguity of position, however, Fanon "diagnosed this general pain as a bona fide symptom" (176).

20. Fanon does not yet diagnose his immigrant, minority patients with a neurological condition. But he does point to the scientific research in colloquia proving the existence of nonspecific somatic expressions of illness. See "The North African Syndrome," 12.

21. Fanon uses the term "neurotic condition" in his writings even as he divests it of its sexual etiology in contexts of extreme relational violence. In his general critique of Freud in colonial contexts, Fanon also implies the rejection of the "war neurosis." Instead, his reading of Freud in this respect suggests an affiliation with Malabou's contemporary assessment that draws on neuropathology to cite a definitive causal relation between war trauma and posttraumatic stress.

22. Ibid., 69. Fanon's metaphor of the anemic personality might have been the effect of a preoccupation with his own blood count. Razanajao et al. note that Fanon regularly monitored his blood count while at Saint Alban's (502).

23. Ibid., 92

24. Ibid., 96.

25. Ibid., 92.

26. Ibid., 92. As stated earlier, in *Toward an African Revolution: Political Essays*, trans. Haakon Chevalier (New York: Grove Press, 1967), Fanon concludes, "The events in Algeria are the logical consequence of an abortive attempt to decerebralize a people" (53). Even if he does not deepen the concept of brain damage in relation to oppression, Fanon nevertheless indicates that inflicting neurological damage on the colonized could result in a trauma with the potential to ignite an explosive situation of revolutionary violence.

27. Ibid., 92.

28. Ibid., 89.

29. Ibid., 94.
30. Ibid., 92.
31. Ibid., 97.
32. Ibid., 94.
33. Ibid., 119.
34. Ibid., 97.
35. Ibid., 96.
36. Brenna Bhandar and Jonathan Goldberg-Hiller in their "Introduction: Staging Encounters" to *Plastic Materialities: Politics, Legality, and Metamorphosis* (Durham, NC: Duke University Press, 2015) write that this is "a passage that seared itself on the brains of many Marxist anticolonial intellectual migrants in New York, Oxford, London, and similar places in the 1960s and 1970s, largely because it often mirrored their own experiences upon arrival in the 'metropole'" (84). Its popularity among members of the intellectual class who felt a visceral link to it could also be attributed to the way it poeticizes the disruptive temporality at work within the experience of expatriation.
37. Fanon, *Black Skin, White Masks*, 89.
38. Ibid., 119.
39. Malabou, *The New Wounded*, 154.
40. Ibid., 10.
41. Ibid., 10.
42. Malabou's argument that destructive plasticity irreversibly ruptures the affective relation has been critiqued for absolving the analyst's responsibility to incorporate the traumatic past into the bounded memory of the social narrative. See Jennifer O. Gammage's "Trauma and Historical Witnessing: Hope for Malabou's New Wounded," *The Journal of Speculative Philosophy* 30, no. 3 (2014): 404–13. Also see Peter Gratton for a critique of the blurring between these two types of suffering. His critical reflection on the relational model suggests that Malabou's theory is not uncontroversial.
43. Ibid., 10.
44. Ibid., 10–11.
45. Ibid., 11.
46. Ibid., xviii.
47. Ibid., xviii.
48. Achille Mbembe in *The Post Colony* suggests that the concept of "deep history" permeates African contexts. Accordingly, the histories of slavery and colonialism in the *longue durée* of capitalist modernity haunt today's forms of violence even as their political identity is effaced. More recently, in "Frantz Fanon's *Oeuvres*: A Metamorphic Thought," *Nka: Journal of Contemporary African Art*, no. 32 (Spring 2013): 8–16, Mbembe argues that Fanon's republished *Oeuvres* reveal the contemporaneity of his ideas that now have a metamorphic power helping us to analyze today's forms of violence, which have metabolized colonial histories.
49. Malabou, *The New Wounded*, my italics, 209.
50. See both *The New Wounded* and *What Should We Do with Our Brain?* Malabou's interpretation of the three types of plasticity—reception, donation, and

explosion—that she conceptualizes on the basis of the new neurological research in brain plasticity.

51. Malabou, *The New Wounded*, 17.

52. Fanon, *Black Skin, White Mask, 94.*

53. Ibid., 119.

54. Malabou, *The New Wounded*, 161.

55. Judith Lewis Herman in *Trauma and Recovery: From Domestic Abuse to Political Terror* (New York: Basic Books, 1997) writes: "The features of posttraumatic stress that become most exaggerated in chronically traumatized people are avoidance or constriction. When the victim has been reduced to a goal of simple survival, psychological constriction becomes an essential form of adaptation. This narrowing implies every aspect of life—to relationships, activities, thoughts, memories, emotions, and even sensations. And while this constriction is adaptive in capacity, it also leads to a kind of atrophy in the psychological capacities that have been suppressed and to the overdevelopment of a solitary life" (cited by Malabou in *The New Wounded*, 157).

56. Malabou, *The New Wounded*, 118.

57. Ibid. Malabou writes, "The posttraumatic state is a gaping wound that appears on all the battlefields of contemporary society. A normal reaction to an abnormal situation, it inscribes the enigma of its event upon the global stage" (161). My reading suggests that the posttraumatic state described by Malabou in today's global context has colonial and inheritance.

58. Sociologist Robert Castel develops the concept of "disaffiliation" in *Metamorphosis de la question sociale* (Paris: Gallimard, 1995). (Cited by Malabou in *The New Wounded: From Neurosis to Brain Damage,* footnote 56, 238.)

59. Malabou, *The New Wounded*, 119.

60. In *The Wretched of the Earth*, trans. *Richard Philcox* (New York: Grove Press, 2004 [1961]), Fanon describes the Manichean relation as a layered geopolitical "reality" *and* psychoaffective trauma.

61. In their "Introduction: Staging Encounters" to their edited collection *Plastic Materialities*, Bhandar and Goldberg-Hiller write that Malabou's plastic subject in *The Future of Hegel* as the subject that "is in some sense structured by its capacity to temporalize itself" (22). But they suggest that this temporalization might not be sufficiently plastic for indigenous justice. They question whether in some situations of oppression the formative plastic powers have the chance to form a different future.

62. Malabou in *The New Wounded: From Neurosis to Brain Damage* cites the American Psychiatric Association. *Diagnostic and Statistical Manual of Mental Disorders*, 3rd edition (DSM-III). Washington, DC: American Psychiatric Association, 1980; revised 3rd edition (DSM-III-R), 1987; 4th edition (DSM-IV), 1994.

63. Malabou, *The New Wounded*, 155.

64. Ibid., 154.

65. Ibid., 150.

66. Ibid., 160.

67. Malabou, *The New Wounded*, 158.

68. Ibid., 155.

69. Ibid., 155.
70. Ibid., 155–56.

BIBLIOGRAPHY

Bégué, Jean-Michel. "French Psychiatry in Algeria (1830–1962): From Colonial to Transcultural." *History of Psychiatry* 7, no. 28 (1996): 533–48.

Bhabha, Homi K. "Forward: Framing Fanon." In *The Wretched of the Earth*, translated by Richard Philcox, vii–xli. New York: Grove Press, 2004.

Bhandar, Brenna, and Jonathan Goldberg-Hiller. "Interview with Catherine Malabou." In *Plastic Materialities: Politics, Legality, and Metamorphosis in the Work of Catherine Malabou*, edited by Brenna Bhandar and Jonathan Goldberg-Hiller, 287–99. Durham, NC: Duke University Press, 2015.

———. "Introduction: Staging Encounters." In *Plastic Materialities: Politics, Legality, and Metamorphosis in the Work of Catherine Malabou*, edited by Brenna Bhandar and Johnathan Goldberg-Hiller, 1–33. Durham, NC: Duke University Press, 2015.

Derrida, Jacques. "Autoimmunity: Real and Symbolic Suicides: A Dialogue with Jacques Derrida. Interview by Giovanna Borradori. *Philosophy in a Time of Terror: Dialogues with Jurgen Habermas and Jacques Derridge*. Chicago: University of Chicago Press, 2003.

Fanon, Frantz. *Black Skin, White Masks*. Translated by Richard Philcox. New York: Grove Press, 2008 [1952].

———. "The North African Syndrome." In *Toward the African Revolution*, translated by Haakon Chevalier, 3–16. New York: Monthly Review Press, 1967.

———. *Le colonisé en question: Le "syndrome nord africain" Pour la révolution africaine: Écrits politiques*, 9–21. Paris: Francois Maspero, 1969.

———. *Toward an African Revolution: Political Essays*. Translated by Haakon Chevalier. New York: Grove Press, 1967.

———. *The Wretched of the Earth*. Translated by Richard Philcox. New York: Grove Press, 2004 [1961].

Gammage, Jennifer O. "Trauma and Historical Witnessing: Hope for Malabou's New Wounded." *The Journal of Speculative Philosophy* 30, no. 3 (2014): 404–13.

Gratton, Peter. "Malabou's Plasticity of the Real." In *Speculative Realism: Problems and Prospects*, 183–200. New York: Bloomsbury Academic, 2014.

Khalfa, Jean. "Fanon and Psychiatry." *Nottingham French Studies*, no. 1 (2015): 52–71.

Khanna, Ranjana. *Dark Continents: Psychoanalysis and Colonialism*. Durham, NC: Duke University Press, 2003.

Macey, David. "Frantz Fanon 1925–1961." *History of Psychiatry* 7, no. 28 (1996): 489–97.

Malabou, Catherine. *The New Wounded: From Neurosis to Brain Damage*. Translated by Stephen Miller. New York: Fordham University Press, 2012.

———. *What Should We Do with Our Brain?* 1st ed. Translated by Sebastian Rand. New York: Fordham University Press, 2008.

Mignolo, Walter D. "Delinking: The Rhetoric of Modernity, the Logic of Coloniality and the Grammar of De-coloniality." In *Globalization and the Decolonial Option*, edited by Walter D. Mignolo and Arturo Escobar, 303–68. New York: Routledge, 2010.

Mbembe, Achille. "Frantz Fanon's *Oeuvres*: A Metamorphic Thought." *Nka: Journal of Contemporary African Art*, no. 32 (Spring 2013): 8–16.

———. *On the Post Colony* (*Studies on the History of Society and Culture*). Berkeley: University of California Press, 2001.

Quijano, Anibal. "Coloniality and Modernity/Rationality." In *Globalization and the Decolonial Option*, edited by Walter D. Mignolo and Arturo Escobar, 22–32. New York: Routledge, 2010.

Razanajao, C. L., J. Postel, and D. F. Allen. "The Life and Psychiatric Work of Frantz Fanon." *History of Psychiatry*, no. vii (1996): 499–524.

CONCLUSION

Chapter 15

Discontinuity and Difference

Heidegger and Lévi-Strauss[1]

Catherine Malabou
(Translated by William Samson)

As a conclusion to the wonderful adventure that is this book, I would like to propose a philosophical hybridization in the form of a confrontation between Heidegger and Lévi-Strauss. It is a strange pairing, since the two authors, while contemporaries, never wrote on one another, even though they overlap on so many points—in particular, and most obviously, on the question of structure. This concept is central to each of them, and yet no dialogue on the subject took place. Heidegger intends to show that structure exclusively designates the temporal mode of being of *Dasein*, while Lévi-Strauss cuts structure off from all existential content. However, in both cases, structure exceeds the context which each of the two thinkers assign to it. On the one hand, as we shall see, structure makes manifest the impossibility of a *simple* division into being and beings, contrary to what Heidegger affirms. On the other hand, structure's ontological anchoring means there can be no *simple* division between the existential analytic and the human sciences, as Lévi-Strauss would have it. What, then, is a structure? In the first place, it designates a mode of articulation, of attachment, of synthesis. But what it binds together is at the same time a play of difference, of detached pieces. Between attachment and detachment, what does structure do? It is this question that will lead me to characterize the *form* of the present volume.

The concept of structure is very present in *Being and Time*. Where does it come from? Heidegger certainly does not borrow it from psychology, nor from sociology (the term, let us remember, is nonetheless adopted by Durkheim in *The Rules of Sociological Method* in 1895), and even less from linguistics. Heidegger uses *Struktur* from his earliest writings on logic, notably as early as 1912, in his *Neuere Forschungen über Logik*.[2] Logic is defined as the science of structure or of valid forms of signification. And the first of these forms is the structure of judgment. Heidegger shows, in his

1925–1926 seminar, titled *Logik: Die Frage nach der Wahrheit*, that the task of traditional logic, that is, of Aristotelian logic, consists of unlocking the fundamental structure of logos (*Grundstuktur der Logos*), which is that of predication, characterized as *Als-Struktur*, or the "as-structure."[3] Certainly, in *Being and Time*, structure is cut off from this logical meaning, but it is nonetheless a product of it.

Without a doubt, Heidegger was not indifferent to the fate that the concept of structure had begun to know in the human sciences. Nor could he ignore the arrival and triumph of structuralism, if only because of his friendship with Lacan. And yet it remains the case that he never spoke of it and never took the trouble to clarify his own understanding of structure vis-à-vis that of the "structuralists." Only twenty years separate the publication of *Being and Tim* and that of Lévi-Strauss's first works: *Family and Social Life of the Nambikwara Indians* in 1948, and *The Elementary Structures of Kinship* in 1949. *Structural Anthropology* groups together texts spanning from 1945 to 1954. And yet there's no exchange between the two authors. Are there nonetheless waypoints—even if they might at the same time be breaking points—between the Heideggerian and structuralist concepts of structure?

The presence of "structure" in the lexicon of *Being and Time* show that it touches on the essential, which is to say, being. More precisely, it touches on that being that has an understanding of its own being, *Dasein*. *Dasein*, moreover, is nothing but a structure: §39 is titled "The Question of the Primordial Totality of *Dasein*'s Structural Whole" (*die Ganzheit des Strukturganzen des Daseins*).[4] *Dasein* understands itself at two levels that interpenetrate and cut into each other. Firstly, when he presents the task and aim of the existential analytic, Heidegger shows that its goal is not to study individual behavior, but to thematically explicate the "ontological structure of *Dasein*." Secondly, in determining this last as a structure of existence (*Existenzstruktur*), Heidegger has brought to light in the same paragraph (§9), the categories which he calls "existentials." Yet the existentials are themselves structures, "highly charged" insofar as they are ontological characteristics of *Dasein*.[5] The existentials, understood as "structural moments," appear as fundamental ontological characteristics of *Dasein*: existentiality, facticity, and fallenness.

What is the relationship between the existential understanding and the logical understanding of structure? In his 1925–1926 seminar, Heidegger showed that the structure of the "as," the structure of predication that Aristotle took to be fundamental, is not, in reality, originary. It refers back to a "signifying foundation . . . anterior to linguistic discursivity. [It] belongs to the sphere, beyond logic, of *Verhalten*—comportment."[6] This comportment, which is in fact the totality of all comportments of human life, is made possible by a primary structure: the prelogical openness, or the "opening" character, of the relationship between *Dasein* and the world.

Sections 39–41 of *Being and Time* do much to clarify that ontological signification, by revealing the unity of the existential in three moments. The first name given to that unity is being-in-the-world. Second, it is called care, a "unitary phenomenon" made in some way apparent by being-in-the-world.[7] Finally, care sees itself driven toward an ultimate unity, temporality: "When we first fixed upon this articulated structure, we suggested that with regard to this articulation the ontological question must be pursued still further back until the unity of the totality of this structural manifoldness has been laid bare. The primordial unity of the structure of care lies in temporality."[8]

Why is the method of *Being and Time* dictated by the clearing of an always higher, truly primordial structural whole? This movement conforms to, among other motifs, the need to divert the structure of that signification which Heidegger calls its ontic signification. This ontic signification is that of an assemblage, an arrangement, a framework, a composition; that is, a machine.

Dasein is a whole, not an aggregate or an assemblage: "Being-in-the-world," Heidegger tells us, "is a structure which is primordially and constantly whole." Certainly, one must insist on the relational and nonsubstantial character of the cooriginary, constitutive moments of wholeness: each can only make sense through its relationship with the others. Be that as it may, that relation deploys itself without ever breaking itself apart, disjuncting, or spacing itself out, contrary to what might take place in an assemblage.[9]

"Is it possible," Heidegger asks, "to come to understand this structured whole . . . of *Dasein* in its wholeness?" The answer is clear: "To put it negatively, it is beyond question that the totality of the structural whole is not to be reached by building it up out of elements. For this we would need an architect's plan."[10] In the same way that he will distinguish, in his *Schelling* (1936), between two meanings of the concept of system, an "authentic" one—"the internal joining which gives a thing . . . its foundation and its tenor," and an "inauthentic" one—"the simply external throwing-together,"[11] Heidegger distinguishes in *Being and Time* between two meanings of *structure* which retread, by the way, almost identically the meanings of *system*. On the one hand, structure as looking "all the way through [the] whole to a single primordially unitary phenomenon which is already in this whole in such a way that it provides the ontological foundation for each structural item in its structural possibility."[12] On the other hand, structure as simple plan, destined to be concretized or fulfilled by an assembly or aggregation of detachable pieces.

Heidegger makes implicit reference to the human sciences—in particular, to anthropology and ethnology—when he enunciates this second meaning of structure. "*Dasein*'s being," Heidegger takes care to clarify, "is not to be deduced from an idea of man" and the structure sought here can in no way issue from a search "directed towards . . . entities within-the-world."[13] As such, "The existential analytic of *Dasein* comes before any psychology or anthropology, and certainly before any biology."[14]

However, worryingly, Heidegger very clearly affirms on the other hand that none of these sciences, nor even the science of logic, directly mobilizes the ontic meaning of structure, the simple assemblage or accumulation of bits and pieces. In section 10, titled "How the Analytic of *Dasein* Is to Be Distinguished from Anthropology, Psychology, and Biology," Heidegger makes an accounting of the work of Dilthey and Scheler. From Dilthey, he mentions "psychology as science of spirit," and from Scheler, he mentions the "phenomenological interpretation of personality." He writes: "The researches of Wilhelm Dilthey were stimulated by the perennial question of 'life.' Starting from 'life' itself as a whole, he tried to understand its 'experiences' in their structural and developmental interconnections."[15] Heidegger adds that Dilthey's "psychology as science of spirit" does not seek to orient itself toward mental elements or psychic atoms, and refuses to make the life of the soul into an aggregate of pieces, preferring to have as an end "life as a whole" and the shapes that it takes.

We see clearly that it is very well structured in the sense of wholeness, of an articulated totality, which is required here, not mere juxtaposition. Similarly, for Scheler, the person appears as a "performance of intentional acts" that "are bound together by the unity of a meaning."[16] Thus, there is no assemblage here, either. As for biology, does it not also give to concepts of "form" or "structure" of the living creature the sense of an organized totality, irreducible to a sum or compilation? Heidegger evidently admits as much when he cites von Uexkull in *The Fundamental Concepts of Metaphysics*: The organism, as permanent structure or form of the animal, "not only sustains its specific unity but gives this unity to itself for the first time."[17]

Why, then, does Heidegger dismiss the ontic sense of structure by relegating it to fields of study from which he himself recognizes that it is absent?

I propose the following answer: *the ontic sense of structure has no meaning*. It does not occur to anyone, not to philosophers, nor to biologists, nor to psychologists, anthropologists, or ethnologists. No one. Not even the man on the street. Certainly, there is some vague, common sense of the word "structure." But by "structure," everyone means the entirety or the articulated unity of a whole. Someone who buys a car that is said, by some conceptual abuse, perhaps, to have a good structure—whether it be said in praise of the chassis or the material the car is made of—does not understand "structure" here to mean a simple aggregate of atoms. Immediately, an approximative idea of organization, rhythm, *synthesis* comes to this person's mind.

What role, then, does this ontic or derived sense of structure play in Heidegger's thought, if this sense in a certain way does not exist and serves no purpose?

Heidegger demonstrates that if the sciences enumerated above cannot give an authentic understanding of *Dasein*, it is because they are not aware of their own ontological foundations. To put it another way, it is because, without being aware of it, they require the ontological sense of structure. It is not because they misuse the term. It is in this way, for example, that if ethnology's "clumsy and crude" way of conceiving things can "be positively helpful in bringing out the ontological structures of phenomena," it is that ethnology "operates with definite preliminary conceptions and interpretations of human *Dasein*."[18] Thus if ethnology is not capable, like the other regional sciences, of bringing to light the totality of *Dasein*, that is, its constitution as a structured whole, in a satisfying way, it is not, once again, because ethnology makes use, implicitly or explicitly, under that name or another, of a derived concept of structure, of an understanding of structure as assemblage, but because ethnology remains blind to the ontological basis of the notion.

Then, if structure always comes back down to an originally synthetic unity, why does Heidegger artificially differentiate it from his sense of the term? Why put an artificial distinction into play, since even for preoccupied *Dasein*, the most degenerate, the least resolute *Dasein* possible, there *still* is nothing that fits the concept of a simple assemblage—nothing, except precisely that which is nothing. Why attempt to break the authentic meaning of structure apart into a false friend (or rather a false enemy) that opposes nothing?

Once more, what purpose does this false, artificial structure of the assemblage serve in *Being and Time*?

Let us leave this question suspended for a time and move to Lévi-Strauss. When he was writing his work on kinship, he had recourse to two very precise conceptions of the notion of structure: one which already had precedence in his discipline, that of a social structure, developed by Anglo-Saxon anthropologists, notably by Radcliffe-Brown; and another, whose discovery came lately to him, which comes from linguistics, and which he discovered through Jakobson's instruction, and by reading de Saussure and Troubetskoy. The originality of Lévi-Strauss's method consists in the way in which he established a link between these two, which allowed him to elaborate a totally new approach to the concept.

For Lévi-Strauss, it must be emphatically affirmed that structure *never* refers to a simple assemblage of elements. Firstly, because, as *Structural Anthropology* makes clear, "From the structuralist point of view, . . . it would be hopeless to try to reach a valid definition of social structure on an inductive basis, by abstracting common elements. . . . If these concepts have a meaning at all, they mean, first, that the notion of structure has a structure."[19] The term "structure" is itself an articulated whole. Secondly, because the relations between different terms of a structure do not form rigid boundaries, but rather permeable thresholds which forbid any rigid separation or parceling out. In a text devoted to the thought of Lévi-Strauss, Jean Pouillon characterizes structure in these terms: "a system of differences that does not lead to their mere juxtaposition, nor to their artificial erasure."[20] Indeed, structure remains at a point of equilibrium, as Lévi-Strauss maintains at the beginning of *The Savage Mind*, between necessity and event, a view from the point of contact between form and phenomenon.

It is useless to emphasize again the obvious connection between the constitutive elements of structure. It is only in bringing together the Heideggerian understanding of articulation (*Gliederung*) with that of Lévi-Strauss that it will become possible to overcome the artificial, ontic signification of structure exposed in *Being and Time*.

In Heidegger, it is a matter of thinking at the same time of difference—the difference between existentials—and unity, which is to say, the fundamental phenomenon which permits different moments to deploy themselves. *Gliederung* is thus inseparable from *Struktur*, and it manifests itself in a certain sense as its synonym, and gives meaning at the same time to the totality formed by the elements of a structure and the particularity of each. Because of this, and regardless of what Heidegger says on the subject, articulation is to be understood here in a quasi-physiological sense. The properly structuralist determination of articulation is also engaged in a necessarily complementary relationship between unity and difference, but it borrows the form of its relation from linguistics, rather than physiology. If, for Lévi-Strauss, it is a matter of designating a type of relation between the structural elements that prioritizes the very existence of these elements, this relation, in order to be systematic, cannot rest on any unitary phenomenon. Fundamental to linguistics, fundamental to social structures and kinship structures is *discontinuity* [*écart*]. Language is certainly a system, which is to say an articulated totality, but that articulation is nothing other than the ordered determination of spacing between distinctive or significant units.[21]

Lévi-Strauss makes discontinuity into a fundamental concept of the structuralist method, as is clearly obvious in *Structural Anthropology*. First, Lévi-Strauss affirms, "What is called a 'culture' is a fragment of humanity which, from the point of view of the research at hand and of the scale on which the latter is carried out, presents significant discontinuities [*des écarts significatifs*] in relation to the rest of humanity. . . . These discontinuities can be reduced to *invariants*, which is the goal of structural analysis."[22] Second: "As in linguistics, it is the *discontinuities* [*écarts différentiels*] which constitute the true subject matter of anthropology."[23]

The question, then, is to see if difference and discontinuity have the same meaning, that is, if Heidegger could have spoken of ontological discontinuity just as easily as of ontological difference. The task is thus to measure the difference or discontinuity between difference and discontinuity. Are difference and discontinuity connected or utterly detached from one another? Heidegger never uses the term "divergence" [*écart*] to refer to the difference between the existentials. Certainly, in *On the Way to Language*, he uses discontinuity [*Auseinander*] to refer to the opening characteristic that implies the proximity between philosophy and poetry. Poetry and philosophy, he says, "diverge from one another entirely."[24] But this discontinuity is not of the same nature as that or those which form and order the intervals between linguistic elements or the relations between relatives in their social circle. Such elements, for the linguist as for the ethnologist, are not properly "neighbors," and the ontological play of their proximity and discontinuity is not in question. This is simply because a discontinuity is just a discontinuity. There's simply a discontinuity, a "hiatus, miniscule and yet invincible," as Foucault tells us in *The Order of Things*, that "resides in the 'and'."[25]

In substituting discontinuity for difference, structuralism tends to invalidate without saying as much, in an unconscious manner, perhaps, the ontological difference that is supposed to be its unperceived foundation. The differential discontinuity has ontological purchase. It is outside of difference even as it operates as pure difference, which is to say it operates as, to take up one of de Saussure's terms, unmotivated difference.

It is now possible to come back to the Heideggerian assemblage while paying attention to the false friend that is structure conceived of as an aggregate. Everything happens as though, through this feint or false sense, Heidegger could conjure in advance what difference would become: a pure discontinuity, nothing but a discontinuity. Indeed, and I've accentuated this point, the system of differential discontinuities is not, for the structuralist, a formal framework where structural elements can accumulate. However, it seems that Heidegger defends himself in advance against a certain interpretation of difference—inadmissible on his account—that would make of difference a discontinuity, and would thus blur the "giving" character of being and beings.

In texts later than *Being and Time*, Heidegger will ceaselessly lay into linguistics, denouncing it as a means of instrumentalizing language. Ceaselessly, he denounces the mechanical, machinic signification that, on his account, the sciences confer on the human, on the human world, and on human language. Even if the assemblage, structure conceived of as an empty shell or formal construction, does not refer to some real entity, it nonetheless has a role: Heidegger uses it, in some way, as an anticipatory shield opposed to the coming of structural difference. It must indeed be said that in France, authors like Deleuze, Foucault, and Derrida, to name only a few, read Heidegger and interpret ontological difference with the help of hermeneutic tools created by structuralism, and notably with reference to the concept of discontinuity [*écart*]. Différance is born precisely out of the encounter of difference and discontinuity. Spatialization and temporalization are more fundamental than the difference between being and beings.

Would Heidegger have anticipated that he would need, in some way, to safeguard structure from structuralism, to protect it from itself, by devaluing discontinuity by reducing it to the interstitial void, the void in the interval that does nothing but separate the elements of an aggregate? For Heidegger, it is still the case that articulation, *Gliederung*, is not a discontinuity. It is certainly suitable to bring this line of thought around articulation back to the really unitary phenomenon that is at its origin: temporality. From the 1925 to 1926 *Logik* seminar, Heidegger shows that time does not operate as *rahenmässig*—as a framework—but *strukturmässig*—as a structure—an unfolding unity articulated by its moments.[26]

One must now see, coming back, that if the notion of discontinuity blurs and worries ontological difference, it is also and precisely because of its own ontological importance. Whatever Lévi-Strauss might say—he, who, from the opening pages of *Tristes Tropiques* in particular explicitly distinguishes his line of thought from that of philosophy and of metaphysics—the properly structuralist determination of structure has an incontestable existential valence. As I pointed to in the introduction of this chapter, things can easily be inverted. Stripping structure of its ontological meaning is itself an ontological gesture.

The structuralist understanding of structure is openness of a certain kind to the finitude that radicalizes being-for-death. It might not be enough to say that *Dasein* is a structured whole. One must moreover show that the relation that unites this whole with others, which is to say other *Dasein*s, is itself structured. Structured by exchange, or by what Lévi-Strauss calls communication. The differential discontinuities or divergences make possible an economy in which each term exists, in one way or another, only in function of its exchangeability. Being finite is to be exchangeable, which is to say having the power and responsibility to circulate. "In any society," Lévi-Strauss writes,

> communication operates on three different levels: communication of women, communication of goods and services, communication of messages. Therefore kinship studies, economics, and linguistics approach the same kinds of problems on different strategic levels and really pertain to the same field. Theoretically at least, it might be said that kinship and marriage rules regulate a fourth type of communication, that of genes between phenotypes. Therefore, it should be kept in mind that culture does not consist exclusively of forms of communication of its own, like language, but also (and perhaps mostly) of *rules* stating how the "games of communication" should be played both on the natural and on the cultural levels.[27]

We can rightfully take structuralism, in this sense, to be pushing the analytic of *Dasein* to its limit in lifting the ban that forces a discontinuity or differentiates discontinuity from difference and throws *Dasein*, perhaps even more violently than *Geworfenheit*, into the economy of the existential marketplace, which even nonexchangeability in the face of death would not end the perpetual motion.

Foucault was not mistaken when, in the final chapter of *The Order of Things*, he saw in the extreme structuralism of the modern *episteme* what he would come to call, in a very unsettling manner, an "analytic of finitude."[28]

Once the obstacle of the artificial meaning of structure as assemblage is done away with, the existential analytic and structuralism can finally really be confronted. On the one hand, because structuralism's account of articulation and of discontinuity can no longer be circumscribed by the limits of a single ontic science or regional ontology, contrary to what Heidegger has stated. It reveals its whole ontological import. By showing itself to be joined together to an analytic of finitude, structuralist structure loses the positivist (i.e., scientistic) ambition that Lévi-Strauss wished to give it.

Moreover, because it hybridizes ontological difference, it crossbreeds Heideggerian purity with inauthenticity. There would, in the end, be no difference or discontinuity between difference and discontinuity.

Being, beings, and their difference, structure speaks all three at the same time. There is no difficulty in describing a being in terms of structure. No matter the being, "be it God, a table, or a washbasin," as Michel Serres once famously said, it can be described as a structure. Being itself, as a fundamentally synthetic whole, is a structure, primordial temporality. As for the difference between them, it appears as the juncture that allows for the articulation of the terms that it both distinguishes and unites at the same time. But what is important is perhaps the discontinuity that allows for that structural synonymy to circulate and to escape empty identity. Without difference, discontinuity is ontologically blind, but without discontinuity, ontology is a tautology. Heidegger and Lévi-Strauss, then, did not agree with each other, but their voices today resonate with the music of a strange duet.

It is in thinking of them that I suggested the title of the present book, *Passionate Detachments*, to the publisher. I was in the process of preparing a course on the concept of structure at the moment of writing this afterword, and it quickly became apparent to me that this concept is at the center of the dialectical relationship between synthesis on the one hand, and separation or fragmentation on the other. Structure is indeed, in the works of these two thinkers, a unifying system and the articulation of particularities independently of one another.

I then thought of the colloquium on my work, organized by Antonio Calcagno and Stephen Lofts. Was it not itself such a structure? Did it not bring together unique, independent reflections that, at the same time, shared something very intense, a *passion*? The themes of this book are obviously varied: philosophy, politics, science, translation theory, psychoanalysis, ecology. Yet they nonetheless come together with the same theoretical care, the same love of thought.

Passionate detachment is precisely the condition of thought. I feel it even more strongly on the occasion of this book, which, while it bears my name, still separates me a little more from “me.”

I would like to thank the organizers of that colloquium, Antonio and Stephen, once more from the bottom of my heart for their hospitality. As well, I’d like to thank CAREP and Western University. A huge thank-you also to all of the participants in the colloquium, who came from so many different places. Finally, I would like to express my sincere gratitude to Isabell Dahms and Thomas Wormald, who gathered together all of these texts and made this book possible. I am deeply indebted to them.

Catherine Malabou
Paris, September 2017

NOTES

1. This is an amended and translated piece previously published in French. Catherine Malabou, “Une différence d'écart: Heidegger et lévi-strauss,” *Revue Philosophique de la France et de l'étranger* 127, no. 4 (2002): 403–16.
2. Martin Heidegger, *Neuere Forschungen über Logik* (1912), *Gesamtausgabe* 1, ed. Friedrich Wilhelm von Herrmann (Frankfort, KY: Klostermann, 1975).
3. Martin Heidegger, *Logik: Die Frage nach der Wahrheit* (Winter 1925–1926), *Gesamtausgabe* 21, ed. W. Biemel (Frankfort, KY: Klostermann, 1976), §12.
4. Martin Heidegger, *Being and Time*, trans. John Macquarrie and Edward Robinson (New York: HarperOne, 1962), 225 [181].
5. Ibid., 70 [44].
6. Jean-François Courtine, “Les ‘Recherches logiques’ de Martin Heidegger, de la théorie du jugement à la vérité de l’être,” in *Heidegger 1919–1929: De l’herméneutique de la facticité à la métaphysique du Dasein*, 7–31 (Paris: J. Vrin, 1996), 25. The 1925–1926 seminar, published as *Logik, die Frage nach der Wahrheit*, is volume 21 of the *GA*.
7. Heidegger, *Being and Time*, 78 [53].
8. Ibid., 375 [327].
9. In his *Zollikoner* seminars, Heidegger reminds us that the existential analytic, much like the Kantian transcendental analytic, does not consist in a dissolution or decomposition of a whole into its elements. It undertakes instead to discover an originary unity, rightly called a structural unity.
10. Heidegger, *Being and Time*, 226 [181].
11. Martin Heidegger, *Schelling: Vom Wesen der menschlichen Freiheit (1809)* (1936), *Gesamtausgabe* 42, ed. I. Schüssler (Frankfort, KY: Klostermann, 1988), 45.
12. Heidegger, *Being and Time*, 226 [181].
13. Ibid., 226 [181–82].

14. Ibid., 71 [45].
15. Ibid., 72 [47].
16. Ibid., 73 [48].
17. Martin Heidegger, *Fundamental Concepts of Metaphysics: World, Finitude, Solitude*, trans. William McNeill and Nicholas Walker (Bloomington: Indiana University Press, 2008), 235.
18. Heidegger, *Being and Time*, 76 [51].
19. Claude Lévi-Strauss, *Structural Anthropology*, trans. Claire Jacobson (New York: Basic Books, 1963), 278.
20. Jean Pouillon, "L'Oeuvre de Claude Lévi-Strauss," *Les Temps Modernes*, no. 126 (July 1956).
21. As Saussure declares: "In Latin, *articulus* means a member, part, or subdivision of a sequence; applied to speech, articulation designates either the subdivision of a spoken chain into syllables or the subdivision of the chain of meanings into significant units; *gegliederte Sprache* is used in the second sense in German." Ferdinand de Saussure, *Course in General Linguistics*, trans. Wade Baskin (New York: The Philosophical Library, 1959), 11.
22. Lévi-Strauss, *Structural Anthropology*, 295
23. Ibid., 328.
24. Martin Heidegger, *Der Weg zur Sprache* (1959), *Gesamtausgabe* 12, ed. Friedrich-Wilhelm von Herrmann (Frankfort, KY: Klostermann, 1985).
25. Michel Foucault. *The Order of Things* (New York: Routledge, 1989), 370.
26. Heidegger, *Logik*, *GA* 21, 409.
27. Lévi-Strauss, *Structural Anthropology*, 296.
28. Foucault, *Order of Things*, 344.

BIBLIOGRAPHY

Courtine, Jean-Francois. "Les 'Recherches logiques' de Martin Heidegger, de la théorie du jugement à la vérité de l'être." In *Heidegger 1919–1929: De l'herméneutique de la facticité à la métaphysique du Dasein*, edited by Jean-Francois Courtine, 7–31. Paris: J. Vrin, 1996.

Heidegger, Martin. *Being and Time*. Translated by John Macquarrie and Edward Robinson. New York: HarperOne, 1962.

———. *Der Weg zur Sprache* (1959). *Gesamtausgabe* 12. Edited by Friedrich-Wilhelm von Herrmann. Frankfort, KY: Klostermann, 1985.

———. *Fundamental Concepts of Metaphysics: World, Finitude, Solitude*. Translated by William McNeill and Nicholas Walker. Bloomington: Indiana University Press, 2008.

———. *Logik: Die Frage nach der Wahrheit* (Winter 1925–1926). *Gesamtausgabe* 21. Edited by W. Biemel. Frankfort, KY: Klostermann, 1976.

———. *Neuere Forschungen über Logik* (1912). *Gesamtausgabe* 1. Edited by Friedrich-Wilhelm von Herrmann. Frankfort, KY: Klostermann, 1975.

———. *Schelling: Vom Wesen der menschlichen Freiheit (1809)* (1936). *Gesamtausgabe* 42. Edited by I. Schüssler. Frankfort, KY: Klostermann, 1988.

Foucault, Michel. *The Order of Things*. New York: Routledge, 1989.

Lévi-Strauss, Claude. *Structural Anthropology*. Translated by Claire Jacobson. New York: Basic Books, 1963.

Pouillon, Jean. "L'Oeuvre de Claude Lévi-Strauss." *Les Temps Modernes*, no. 126 (July 1956).

Saussure, Ferdinand de. *Course in General Linguistics*. Translated by Wade Baskin. New York: The Philosophical Library, 1959.

Index

About the Contributors

Andrew Bevan is a PhD candidate at the Centre for Research in Modern European Philosophy at Kingston University in London, UK. He holds an MA in modern European philosophy and MSc in cognitive science. His research centers on the concept of affect and its relation to time as it functions in philosophy and neuroscience with the aim of connecting the affective turn with new materialisms.

Isabell Dahms is a PhD candidate at the Centre for Research in Modern European Philosophy at Kingston University in London, UK. Her doctoral research explores the concept of speculation in eighteenth-century German philosophy, particularly Kant and Hegel, and the contemporary relevance of this concept for feminist analysis and critique, with special focus on the work of Judith Butler and Luce Irigaray.

Sujaya Dhanvantari is a PhD Candidate at the Centre for Interdisciplinary Studies in Society and Culture at Concordia University. Her research focuses on the intersections of continental philosophy, critical race theory, feminism, and postcolonial studies. Her PhD dissertation *Anticolonial Philosophy (1947-1961): Simone de Beauvoir & Frantz Fanon* views Beauvoir's concept of "situated" ethical freedom in *The Ethics of Ambiguity* as a critical inheritance for Fanon's anticolonial texts, *Black Skin, White Masks* and *The Wretched of the Earth.*

Cristóbal Durán is associate professor at the University Andrés Bello in Chile. He is also a researcher of FONDECYT, and is currently working on a research project about the ontology of relations and the theory of multiplicities on the philosophies of Simondon and Deleuze. He has published several articles and essays on Derrida, Deleuze, Malabou, Simondon, and aesthetics, and three books: *Temblores, del cuerpo sonoro de Hegel* (2014), *Amor de la música: Patricio Marchant* (2016), and *Polifonías: Jacques Derrida, la voz, la sorpresa* (2017). Currently, he is finishing two books: one about the cinema of David Cronenberg, and another about the question of the body in Malabou's philosophy.

Deborah Goldgaber is assistant professor of philosophy at Louisiana State University. She is currently completing a book manuscript connecting her account of deconstructive materialism to contemporary debates in speculative realism and new materialism. The project, entitled *Speculative Grammatology: Derrida and the New Materialism*, is contracted with Edinburgh University Press. She has articles forthcoming in *Journal of Speculative Philosophy*, *Postmodern Culture*, and *The Routledge Handbook of Philosophy and Translation.*

Sandrine Hansen is a PhD candidate at the Institute of Philosophy at KU Leuven in Belgium. Her research project entitled "Habituating Selves: An Investigation of the Limits of Plasticity" is funded by the Research Foundation – Flanders (FWO). She is the cofounder of "selskab for filosofisk feminisme" [Society for Philosophical Feminism] based in Copenhagen.

Catherine Malabou is professor at the Centre for Research in Modern European Philosophy at Kingston University in London, UK, and a professor of comparative literature and European languages and studies at the University of California, Irvine. She is the author of *Before Tomorrow: Epigenesis and Rationality* (2016, trans. Carolyn Shread); *Self and Emotional Life: Merging Philosophy, Psychoanalysis, and Neuroscience* (with Adrian Johnston, 2013); *The Ontology of the Accident: An Essay on Destructive Plasticity* (2012, trans. Carolyn Shread); *The New Wounded: From Neurosis to Brain Damage* (2012); *The Heidegger Change: On the Fantastic in Philosophy* (2012, trans. Peter Skafish); *Changing Difference: The Feminine and the Question of Philosophy* (2011, trans. Carolyn Shread); *Plasticity at the Dusk of Writing: Dialectic, Destruction, Deconstruction* (2009, trans. Carolyn Shread); *What Should We Do with Our Brain?* (2009, trans. Sebastian Rand); and the forthcoming *Metamorphoses of Intelligence: What Should We Do with Their Blue Brain?* (trans. Carolyn Shread).

Meadhbh Mcnutt is a writer and artist based in London, UK. Her research interests include critical theory, neuroscience, politics, and contemporary art. Since graduating from the Centre for Research in Modern European Philosophy at Kingston University, Meadhbh has worked in gallery and editorial assistance and is a regular contributor to *Tank* magazine. She has exhibited artwork in the United Kingdom, Ireland, Hungary, Poland, and Hong Kong.

Georgia Mouroutsou has been an assistant professor of philosophy at King's University College and a member of the Graduate School for Ancient Philosophy at Western University since 2014, following two postdoctoral fellowships at Cambridge, UK, and Humboldt Berlin, Germany. She has published a monograph and a number of articles on later Platonic metaphysics and ethics. She has also developed a strong interest for plasticity, time, and affectivity in the later Stoics, especially Marcus Aurelius. Her current book project concerns Plato's double dialectic of pleasure, namely his critical dialogue with different types of hedonists and his metaphysical analysis of various kinds of pleasure.

Nancy D. Nisbet works in the Schools of Visual Art and Critical Creative Studies at the Alberta College of Art and Design. Her transdisciplinary qualifications and research blurs the boundaries between philosophy (PhD), contemporary art (MFA), and molecular genetics (BSc). Recent publications include "Conceiving a Subject of Mutation: Event, Plasticity, and Molecular Worlds" and "Repenser Exchange: événement, idéologie et procédure artistique." Her nationally-funded artworks have been exhibited and presented in the Netherlands, Germany, Japan, Thailand, Argentina, Mexico, the United States, and Canada. She is currently working on a transcendental materialist project that rereads Merleau-Ponty's *Phenomenology of Perception* together with empirical research on molecular communication.

John Nyman is a PhD candidate at the Centre for the Study of Theory and Criticism at Western University in London, Ontario. His research focuses on Jacques Derrida's writing under erasure, deconstruction, and contemporary experimental poetry including the emerging genre of erasure poetry. John is also a poet working in visual, conceptual, and traditional verse forms; his debut collection, *Players*, was shortlisted for the 2016 Gerald Lampert Memorial Award.

Joshua Schuster is associate professor of English at Western University. His first book is *The Ecology of Modernism: American Environments and Avant-Garde Poetics* (2015). His recent essays have appeared in *Parrhesia*, *Resilience: A Journal of the Environmental Humanities*, *Humanimalia*, *Minnesota Review*, and *Photography & Culture*, as well as in the edited volume, *Critical Perspectives on Veganism*, and in the book *Edward Burtynsky: Essential Elements* (2016). He is currently working on the book project, *What Is Extinction? A Cultural and Natural History of Last Animals*.

Carolyn Shread is a lecturer in French at Mount Holyoke College and also teaches translation at Smith College in Massachusetts. Both a scholar and translator, Shread has translated eleven books, including five by French philosopher Catherine Malabou. Most of her published articles address three principal areas of research: Bracha Ettinger's proposal of metramorphosis in the context of feminist translation, the implications of Malabou's concept of plasticity for translation studies, and the process of translating Haitian author Marie Vieux-Chauvet's *Les Rapaces* into English.

Michael Washington is a PhD candidate in philosophy at the Centre for Research in Modern European Philosophy in London, UK. His research focuses on queer studies, with an interest in the ways it intersects with continental philosophy, black studies, and psychoanalysis.

Thomas Wormald is a PhD candidate at the Centre for the Study of Theory and Criticism at Western University in London, Ontario. His research focuses on the thought of Catherine Malabou, with his doctoral project being a historical and philosophical exploration of the concept of plasticity from British seventeenth-century philosophy to German Idealism.

Lightning Source UK Ltd.
Milton Keynes UK
UKHW03n1953170418
321213UK00001B/16/P

9 781786 606921